Reducing Water Use and Carbon Footprint: working toward a circular economy

Reducing Water Use and Carbon Footprint: working toward a circular economy

Edited by
Magdalena Zabochnicka

IWA PUBLISHING

Published by **IWA Publishing**
Unit 104–105, Export Building
1 Clove Crescent
London E14 2BA, UK
Telephone: +44 (0)20 7654 5500
Fax: +44 (0)20 7654 5555
Email: publications@iwap.co.uk
Web: www.iwaponline.com

First published 2026

British Library Cataloguing in Publication Data
A CIP catalogue record for this book is available from the British Library

ISBN: 9781789065121 (paperback)
ISBN: 9781789065138 (eBook)
ISBN: 9781789065145 (ePub)

Doi: 10.2166/9781789065138

This eBook was made Open Access in January 2026.

Contents

About the Editor

Associate Professor Magdalena Zabochnicka, PhD DSc, is a multidisciplinary scientist with an educational background spanning environmental engineering, business English, pharmaceutical sciences and medical analytics. She acts as a researcher and lecturer at the Czestochowa University of Technology, Faculty of Infrastructure and Environment in Poland. She was a fellow of the Polish Ministry of Science and Higher Education's 'Top 500 Innovators' program at the Universities of Oxford and Cambridge. She also acted as a Visiting Professor at the University of Cagliari, Sardinia, in Italy.

Her scientific work focuses on environmental technologies, including water and wastewater treatment, waste management, CO_2 capture, (bio)sorption, bioplastics and renewable energy. As an Innovation Coach and Expert at the Polish Ministry of Development Funds and Regional Policy, she helps boost the competitiveness of Polish enterprises by implementing cutting-edge technologies and sustainable solutions, supported by European funding. Her expertise extends to European initiatives such as the Circular Economy, Green Deal, and Bioeconomy Strategy, and she serves as an expert for the European Commission, Polish Agency for Enterprise Development, and the Polish National Centre for Research and Development in Poland.

A passionate advocate for interdisciplinary thinking and responsible environmental management, Professor Zabochnicka promotes strategies such as life cycle assessment (LCA) and circular waste management (6R). She is the author or co-author of over 150 scientific publications, one patent on production of algal biomass, one patent on wastewater treatment, ten patents on biofoils with fillers from wastes, and one utility model - a photobioreactor for cultivation of algal biomass. For her inventions she has won many international awards, including the Award of 100 Anniversary of the Polish Patent Office. She has participated in more than 40 professional training courses, such as Project Management Prince2 Foundation, Research Data Management (RDM), Intellectual Property Management, Research Commercialization, Innovation Coaching, Mentoring, Tutoring, e-learning, and EU Horizontal Policies. She lectures in both Polish and English, supervises numerous theses, and actively contributes to European R&D projects as a Principal Investigator, coordinator, manager and researcher.

doi: 10.2166/9781789065138_0001

Introduction: Reducing water use and carbon footprint: working toward a circular economy

M. Zabochnicka[1]*

Czestochowa University of Technology, Faculty of Infrastructure and Environment, J.H. Dąbrowskiego 69, 42-201 Czestochowa, Poland
**Corresponding authors: magdalena.zabochnicka@pcz.pl*

The current environmental, economic, and social needs for innovative technological solutions should fit into both the action program of the 2030 Agenda for Sustainable Development and the European Green Deal implementation strategy to reduce water use and our carbon footprint.

In terms of environmental and social sustainable development, the book fits directly into activities aimed at tackling climate change. Implementation of the innovative solutions helps to improve air quality and the management of water. As a result, the quality of life and health of the society can be improved.

In economic terms, the results of the research should attract many industries. The technologies developed should fill the gap in the market by proposing low-cost and effective methods for the purification of exhaust gases and water management. There is a huge market for effective methods for reducing water and carbon footprint by innovative CO_2 capture from many industries such as power plants, combined heat and power plants, coke plants, waste disposal plants and post-process waste management, and working toward the circular economy.

One of the challenges now is to develop strategies to reduce greenhouse gas emissions. Methods to reduce carbon emissions using innovative techniques are essential needs. Strategies for the management of post-process wastes should be provided in accordance with the assumptions of the circular economy. In particular, the 6R concept based on the principles of refuse, reduce, reuse, repair, recycle, and rethink are helpful.

Numerous studies have shown progress in carbon capture by using various chemical reagents for CO_2 utilization such as amines, calcium carbonate, etc. However, much less attention is paid to reduce water use and carbon footprint by utilization of innovative water-soluble sorbents for carbon capture and post-process waste management toward the circular economy.

Reaching net-zero carbon emissions is the main driving force for the development of technologies targeted at CO_2 capture, storage, and utilization, which is the main subject of this book. The book familiarizes readers with selected achievements in the field of the sustainable development economy. Learning from the book could be used in courses for advanced and graduate students and for scientists in research work, as well as for managers in many industries that are looking for solutions to lower their carbon footprint and need for water. Particularly, the book presents the innovative utilization of liquid nanoparticle ferriferrohydrosol (FFH) for CO_2 capture to reduce carbon footprint. Innovative aqueous solution-based sorbent, besides reactions with carbon dioxide, also possesses coagulating and catalytic characteristics. The innovation is also the treatment for, and/or finding new potential application for, postprocess wastewater that can help reduce the water footprint.

In Part 1, innovative carbon capture methods have been described. Chapter 1 presents a comparative analysis of various methods of CO_2 capture from exhaust gases and on developing new ways to recycle post carbon capture wastes in accordance with the aims of the circular economy. Chapters 2 and 3 focus on carbon capture by electrochemically produced FFH as a cheap, available, convenient, and ecologically acceptable reagent. Chapter 4 presents the research of steam gasification of power coal and biomass with catalytic steam conversion of the produced synthesis gas into hydrogen. The authors proved the lowest cost of CO_2 removal in solid fuel conversion processes. In Chapter 5, landfill gas calorific value enhanced by carbon dioxide extraction to boost power generator efficiency is explored. Chapter 6 mainly focuses on the requirements that SMEs have to fulfill regarding CO_2 emissions. The authors stated that industries which emit CO_2 into the environment must obtain an administrative decision regarding the use of the environment, prepare annual emission reports, pay environmental fees, and submit appropriate reports.

In Part 2, the innovative industrial waste and water management technologies are discussed. In chapter 7, the authors prove that the conversion of solid organic wastes and biomass show that the use of a steam plasma torch enables the reduction of the impact of solid organic wastes (organic fermentation products–greenhouse gases CH_4 and CO_2) on the environment and to obtain additional valuable products–bio-char and hydrogen-enriched synthesis gas. Obtained hydrogen-enriched gas can be used for the synthesis of valuable organic compounds. The authors plan to start research using oxygen-enriched air and a steam plasma torch for the operation of the downdraft gasifier. Industrial water used for such processes could be recycled, reused, or reconditioned to lower the water footprint. Chapter 8 investigates the characteristics of sustainable dry-pressed ceramic tiles made by adding fine granite waste (FGW), such as the modulus of rapture (MOR), bending strength (BS), and assessing the quantity of water absorption (WA%).

All of the research achievements are in agreement with the rules of circular economy. The capture of CO_2 from exhaust gases can help to reduce carbon footprint by utilization of such immobilized gases to produce new materials. Moreover, management of post-carbon-capture wastes can enhance the lowering of the water footprint.

Part 1

doi: 10.2166/9781789065138_0005

Chapter 1

Carbon dioxide capture methods and waste management as an element of the circular economy

U. Kępa, M. Worwąg* and M. Zabochnicka**

Czestochowa University of Technology, Faculty of Infrastructure and Environment, J.H. Dąbrowskiego 69, 42-201 Czestochowa, Poland

**Corresponding authors: urszula.kepa@pcz.pl; malgorzata.worwag@pcz.pl; magdalena.zabochnicka@pcz.pl*

ABSTRACT

This chapter presents a comparative analysis of various methods of CO_2 capture from exhaust gases. Methods based on chemical absorption are discussed. The use of substances such as potassium bicarbonate, aqueous ammonia solution and the most commonly used, amine solvents (monoethylamine, diethylamine), is presented. The course of the reaction for individual methods is given, and their effectiveness and costs compared. Their advantages and disadvantages are analyzed.

The chapter also presents the types of waste generated after CO_2 capture and the possibilities for their management. CO_2 capture technologies are one of the most important methods to achieve the EU's climate goals, and it is observed that they are increasingly used in industry: in electricity and heat generation plants, cement plants, and municipal waste thermal treatment facilities. Innovations concern the technologies themselves and their effectiveness, there is little research on the waste generated after the CO_2 capture process. The type and amount of waste generated is closely related to the technology used. According to the European Union's policy on the circular economy, all the processes should aim to reduce the amount of waste generated and propose appropriate methods for its treatment. In order to reduce greenhouse gas emissions by reducing the carbon footprint, waste-free technologies or the use of recycled materials in them should be pursued. Current modifications of the method are moving toward a closed loop, especially in terms of the waste materials used and their recycling that is connected to the circular economy.

Keywords: amine solvents, ammonia, chemical absorption, CO_2 capture, potassium carbonate, waste, waste management, resource recovery, circular economy

1.1 INTRODUCTION

It is currently believed that carbon dioxide is the main factor causing the greenhouse effect. The impact of increasing carbon dioxide concentration in the atmosphere on global warming has been undoubtedly proven. The CO_2 in the atmosphere comes not only from anthropogenic sources, but also from natural ones. However, the impact of the burning of fossil fuels on the currently observed climate anomalies is unquestionable. Carbon dioxide is produced primarily during the combustion of fuels in the processes of generating electricity and heat (Bargieł and Zabochnicka-Świątek, 2018; Zabochnicka, 2022). According to a report prepared by the International Energy Agency (IEA), 36.8 GT (billion tons) of CO_2 was emitted into the atmosphere in 2022. Compared to 2021, when emissions were 36.3 GT, this was the highest ever year-on-year increase of 6% (CO_2 Emissions in 2022, 2023). Emissions from Polish installations amounted to nearly 185 million tons, of which the commercial power sector is responsible for ca. 55%. Compared to emissions in 2021, it decreased by approximately 4% (Ministry of Climate and Environment, Department of Strategy and Analysis, 2022).

Numerous research works on possible methods of CO_2 removal are carried out in many academic and industrial centers (Czech *et al.*, 2020; Krzywonos *et al.*, 2014). They concern both the capture of CO_2 from waste gases and the possibility of removing it directly from the atmosphere. Taking into account the stage of the process at which separation occurs, emission reduction methods can be divided into:

- CO_2 removal from exhaust gases,
- separation of CO_2 from gas fuel before combustion,
- fuel cell processes,
- fuel combustion in an oxygen atmosphere (oxygen-fuel process),
- separation of coal from fuel before combustion (Hydrocarb process).

In the case of capturing carbon dioxide directly from waste gases, the following processes are applicable: absorption, adsorption, membrane separation, and cryogenic methods.

Due to the structure of the power industry in Poland, which is mostly based on coal and uses condenser units, post-combustion methods of removing carbon dioxide from flue gas are among the most important sequestration methods. Absorption is considered the most favorable method for carbon dioxide capture from flue gas. This is due to the high efficiency of the process and the possibility of obtaining a product with a high degree of purity. Once captured, carbon dioxide can be stored (CCS technologies) or used, for example, in the chemical or fuel industry (CCU technologies). The proper management of post-process wastes is always a challenge (Zabochnicka-Świątek, 2013). There are still developing new ways to recycle such wastes to be in accordance with the circular economy.

1.1.1 Advanced amine process

The most commercially used and readily available separation technology involves the absorption of CO_2 in an aqueous solution of amines. Carbon dioxide is then desorbed, dried, compressed, and transported to a storage location or used as a substrate in other processes.

Most commonly, a 30% solution of monoethanolamine (MEA) or methyl diethanolamine (MDEA) is used in the process. The reaction accompanying CO_2 absorption is reversible if the solution is heated to 383 K (Bochon & Chmielniak, 2015):

$$2HO - C_2H_4 - NH_2 + CO_2 + H_2O \rightarrow (HO - C_2H_4 - NH_3)^+ + H_2CO_3^-$$

The process is used at CO_2 concentrations in the range of 3–20%. A removal of 0.4 kg CO_2/kg MEA is achieved. After the absorption process, the used sorbent must be regenerated, which significantly reduces the thermal efficiency of the entire system as significant amounts of heat are required for the desorption process, what's more is that the solvent cannot be fully regenerated. Disadvantages of the method also include the corrosiveness of the reaction environment, the effect of the presence of oxygen in the gases, resulting in the degradation of the absorption liquid, and the negative effect of trace impurities. Furthermore, the flue gas requires desulfurization of less than 5 ppm SO_2 before the absorption process.

For example, the technology is being used to capture carbon dioxide at the Boundary Dam coal-fired power plant in Canada. The full operational capacity of the project was reached at the end of 2016.

1.1.2 Chilled ammonia

The method involves cooling the flue gas to ca. 283 K, condensation of water vapor and reducing the flux, and then CO_2 absorption in a suspension of carbonate and acidic ammonium carbonate. After reacting with the absorbent, ca. 90% of CO_2 forms acidic ammonium carbonate according to the following reaction (Czech *et al.*, 2020):

$$(NH_4)_2CO_3 + CO_2 + H_2O \rightarrow 2NH_4HCO_3$$

The process is used at CO_2 concentrations in the range of 3–20%. A removal of 1.2 kg CO_2/kg NH_3 is achieved. Gases purified in the absorber are further washed to remove residual ammonia before being released into the atmosphere. The post-absorption suspension is heated in a heat exchanger to dissolve the acidic ammonium carbonate, which is decomposed in a heated pressure regenerator at a temperature of 393 K and a pressure of 2 MPa. A stream of compressed pure CO_2 is formed whereas the ammonium carbonate solution is returned without significant loss to the absorber. The process is characterized by low energy consumption, and the presence of oxygen and pollutants in the gases does not reduce the efficiency of CO_2 absorption.

1.1.3 Absorption using potassium carbonate

Potassium carbonate has many advantages as it is inexpensive, environmentally harmless, resistant to degradation, and requires low energy inputs for regeneration. The main problem with its use is the low reaction rate, which is attempted to be intensified by the use of catalysts.

Carbon dioxide capture occurs according to the following reactions (Skurygin & Poroyko, 2020):

$$CO_2 + K_2CO_3 + H_2O \rightarrow 2KHCO_3$$

The capture process is conducted at 298 K, whereas the CO_2 desorption process occurs at 363 K. High-temperature processes are also analyzed to intensify the reaction.

The results presented in the paper (Chuenphanab *et al.*, 2022) indicate that the maximum CO_2 removal efficiency is approximately 87% for the K_2CO_3 solution and approximately 73% for the MEA solution. Comparing the annual cost of CO_2 capture, the results were USD 57.50/t CO_2 for potassium carbonate and USD 107.50/t CO_2 and for MEA.

To achieve specific CO_2 reduction targets, it is necessary to equip power plants burning coal or gas with facilities to reduce carbon dioxide emissions. In the case of the construction of new power units, this will involve an increase in expenditures, at the same time increasing the internal energy consumption. Furthermore, existing units require CCS installations. This will undoubtedly increase energy prices.

It was found that with a full CCS system, the most important item of total costs (capital expenditures and operating costs) is the capture and compression of carbon dioxide from power installations or industrial processes. These costs, depending on the author, are reported to account for 80–85% of the total. Transportation costs usually account for about 10% and storage costs—for 10–15%.

The situation in the European energy market has been changing rapidly in recent years. Currently, the situation is further exacerbated by the ongoing Russo-Ukrainian War with the related sanctions imposed on Russia, and the significant reduction in coal and gas supplies. Over the past two years, there has been a rapid increase in the price of carbon allowances. Until 2017, prices remained stable at less than €10 per ton of CO_2. Over the past few years, future prices have risen significantly and are now about three times higher than they were just two years ago (Figure 1.1).

On the one hand, this increases the price of power generation, but on the other, it may contribute to wider commercial use of carbon capture installations. The current high prices suggest the increasing viability of the construction of CCS or CCU systems using technologies of CO_2 capture from flue gas. Proceeds from carbon allowances on avoided CO_2 emissions can be considered revenues from these projects.

As of today, there are no large operating CCS projects in EU member states. In Europe, such installations are present only in Norway.

When looking for ways and materials used to capture CO_2, an interesting concept is to use solid waste for this purpose. CO_2 capture using waste is accomplished through three paths:

- Using alkaline waste—Due to the acidity of CO_2, alkaline materials such as silicates and oxides rich in calcium and magnesium can react with CO_2 and produce solid carbonate products. This is referred to as CO_2 capture mineralization or carbonate mineralization. Industrial wastes used in CO_2

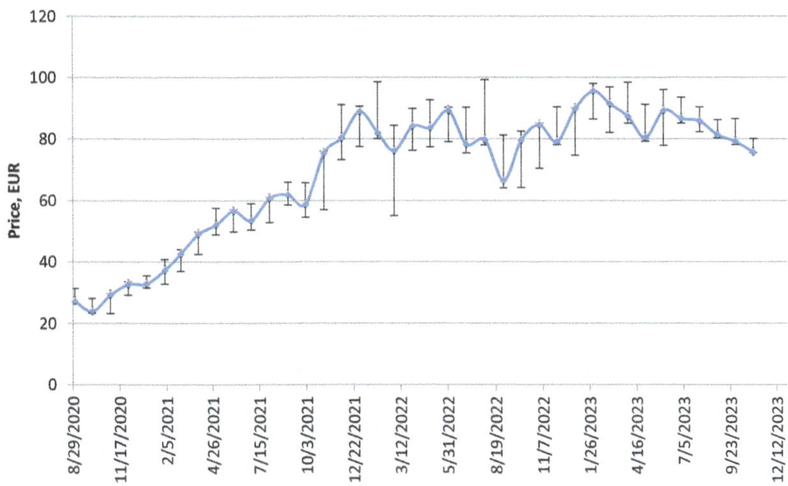

Figure 1.1 Increase in prices of CO_2 emission allowances at EU ETS market.

capture are those that are rich in alkaline bonds: steel slag and carbide slag. The process can be controlled by the mineralization conditions and valuable end products can be produced. Given the quantities and occurrence of solid alkali wastes, mineral carbonation is one of the most potential CO_2 capture technologies that can be applied on a large scale.

- The second section includes examples of CO_2 conversion to valuable chemicals using solid waste-based catalysts solid waste-based catalysts. In addition to CO_2 carbonization, solid waste can be used as catalysts in the conversion of CO_2 to valuable chemicals. Since solid wastes are cost-effective and easy to recycle, cost-effective CO_2 conversion routes can always be developed. Wastes used in this method include: bio-based wastes, red mud, fly ash, sludge, petroleum coke.
- The third section focuses on pyrolysis of solid waste based on carbon to syngas and biocarbon. Wastes used for pyrolysis are, for example, sesame seed waste, tea waste, oak sawdust, orange peel, and lipid wastes. Biomass pyrolysis is an efficient way to generate syngas, which can be used to produce biofuels or other useful chemicals (Cui & Kær, 2020; Foong *et al.*, 2020). Pyrolysis of biomass wastes under a CO_2 atmosphere allows full utilization of the waste. The gaseous products can be used as fuel or intermediates for chemical synthesis, biocarbon is also a useful material that can be used as a catalyst in processes and biodiesel production, and so on. Carbon-based pyrolysis of solid waste with CO_2 still faces several key problems. Among which the most important are high energy consumption. Besides, selectivity to target products and separation of gaseous and liquid products also need to be considered. Therefore, an energy balance should be carried out, and product distribution should be for positive environmental benefits.

Considering the increasing amount of solid waste that poses a threat to the environment, CO_2 capture and utilization through its mediation cleverly combine the disposal of these two types of waste. Interestingly, the easy availability of most solid waste makes large-scale CO_2 capture and utilization feasible and cost-effective. The current annual discharge of solid waste is about 1.3 billion tons, and it is estimated that this figure will approach 40 billion tons by 2050 (Khandelwal *et al.*, 2019). The use of solid waste from agriculture, coal, metallurgy, and the household sector is also a public concern. Therefore, it is urgent to find a suitable disposal method for these solid wastes. To date, tremendous efforts have been made in the direction of solid waste utilization and technologies such as pyrolysis, dehydration, vacuum metallurgical separation, and hydrothermal technology have been developed (Anuar Sharuddin *et al.*, 2016; Hong *et al.*, 2015; Zhan & Xu, 2014). More recently, solid waste has found application in CO_2 capture and utilization, serving as a material for and activation material or co-transformation with CO_2. From the perspective of CO_2 transformation and solid waste utilization, this strategy represents an exciting prospect. Solid wastes such as fly ash, biomass waste, sewage sludge, and so on have a high capacity to capture CO_2 when modified with amines. There are numerous reports on the use of solid waste loaded with waste amines for CO_2 capture (Itakura *et al.*, 2006; Olivares-Marín & Maroto-Valer, 2012; Shen *et al.*, 2021a, 2021b; Wang *et al.*, 2017). Adsorption of CO_2 after conversion with amines provides great potential for in situ CO_2 capture.

The use of waste materials in CO_2 capture technologies seems attractive, but they face a number of problems, such as:

- difficulties in implementing the technological scale of CO_2 mineralization with alkaline wastes due to the lack of standards for construction materials from carbonization,
- the cost of carbonization of wastes such as fly ash, steelmaking slag and carbide slag is high. It is more economically advantageous to use this type of solid waste in other sectors,
- from a life-cycle perspective, the further use of spent solid waste-based catalyst should be explored,
- the difficulty and complexity of the processes generates costs for the method of CO_2 capture using solid waste, so cheaper technological solutions should be sought to make the balance of CO_2 capture favorable compared to the costs incurred.

Regardless of the challenges, CO_2 capture and utilization through solid waste is still considered a promising strategy for dealing with CO_2 emissions and solid waste accumulation. The hope is that by considering the life cycle of the material and designing sensible CO_2 capture and utilization, carbon-neutral and even carbon-negative technologies can be developed.

In the search for materials capable of capturing or chemically transforming CO_2, in line with recent trends toward a closed-loop materials economy, more emphasis is being placed on functional material design and production from societal waste. Porous materials such as polymer adsorbents and metal-organic structures (MOFs), which have had considerable success in capturing and utilizing carbon dioxide, have traditionally been synthesized from non-renewable petrochemical feedstocks. Currently, the search for materials used in CO_2 capture technologies is exploring the possibility of using waste plastics as a hitherto overlooked low-cost alternative resource for synthesizing functional materials for CO_2 capture and utilization. Such a solution not only reduces the amount of discarded plastic waste, but also offers reuse of these polymers, keeping them in continuous use. Work on the application of polymers in CO_2 capture, focuses on how to create and transform the structure of polymers to obtain their high porosity and thus their effectiveness in technology. The biggest problem in the alternative materials used for CO_2 capture is technological cost, despite their high efficiency, the proposed technologies need to be modified to reduce costs. While polymeric adsorbents made from waste plastics allow a high degree of bottom-up design for selective and high CO_2 capture, their often relatively complex synthesis can make scaling difficult. Porous carbon materials generally have high physicochemical stability, but their CO_2 selectivity and adsorption capacities are often poor. Although polymeric adsorbents can provide CO_2 selectivity, high adsorption capacity and are versatile in applications (i.e., they also enable CO_2 utilization), they can be difficult and expensive to produce on a large scale. Indeed, there are many opportunities for further exploration, as only PET, PVC, PU, and PS are currently suitable for conversion to CCU materials, and up to 70% of the mass of available waste plastics remains to be used. Recycling polymers to produce CCU materials for CO_2 capture can significantly reduce the cost of producing the materials, with the added benefit of reducing the use of petroleum-based raw materials for their production. The structural diversity of waste plastics offers a wide range of innovative materials for CCU. In addition to the most common plastics, niche polymers, for example engineering thermoplastics such as polyetheretherketone (PEEK), can also be considered as starting materials that provide excellent mechanical strength. It is even possible to use next-generation bio-based plastics, such as polyhydroxyalkanoates. In addition, the possibility of using plastic-derived materials to directly capture CO_2 from the air should be considered, which will be a significant breakthrough for both achieving a closed-loop materials economy and a zero-carbon society (Teo *et al.*, 2022). Management of post-process wastes is usually a part of CO_2 capture technologies. Such an approach fulfils the aims of the circular economy.

1.2 CONCLUSIONS

(1) CO_2 capture using absorption is a process that is economically and technologically efficient and industrially viable.
(2) Absorption in amine solution is among the most commercially used absorption methods of carbon dioxide capture in post-combustion processes.
(3) Current prices for CO_2 allowances under the European Union Emissions Trading System (EU ETS) are close to the global costs of CCS.

(4) The increasing CO_2 emissions prices should encourage the design and construction of CSS installations in the future, both for new projects and existing plants.
(5) In CO_2 capture technology, closed-loop economics should be taken into account, including the adsorbent materials used in the process.
(6) Waste materials, after appropriate treatment, can be used in CO_2 capture technologies.
(7) Research work focuses on the search for cheap and readily available waste materials and the application of a closed-loop CO_2 capture method.
(8) Plastic-derived waste is readily available and can be used to produce CCU materials for CO_2 capture, with the dual benefits of waste recycling and CO_2 capture.
(9) Plastic-derived materials can be used to directly capture CO_2 from the air, which will be a significant breakthrough for both achieving a circular economy and a zero-carbon society.
(10) The biggest problem in the proposed innovative solutions is the cost of the process, currently striving to optimize the process line from an ecological and economic point of view.

ACKNOWLEDGMENTS

This work was supported in the framework of the EUREKA NETWORK project, "Capture of carbon dioxide by innovative sorbent – InnoCO2sorbent" (EUREKA/InnoCO2Sorbent/2/2021), financed by the National Centre for Research and Development.

REFERENCES

Anuar Sharuddin S. D., Abnisa F., Wan Daud W. M. A. and Aroua M. K. (2016). A review on pyrolysis of plastic wastes. *Energy Conversion and Management*, **115**, 308–326, https://doi.org/10.1016/j.enconman.2016.02.037

Bargieł P. and Zabochnicka-Świątek M. (2018). Technologies of coke wastewater treatment in the frame of legislation in force, *Ochrona Środowiska i Zasobów Naturalnych*, **29**(1), 11–15, https://doi.org/10.2478/oszn-2018-0003

Bochon K. and Chmielniak T. (2015). Energy and economic analysis of the carbon dioxide capture installation with the use of monoethanolamine and ammonia, *Archives of Thermodynamics*, **36**(1), 93–110, https://doi.org/10.1515/aoter-2015-0007

Chuenphanab T., Yurata T., Sema T. and Chalermsinsuwan B. (2022). Techno-economic sensitivity analysis for optimization of carbon dioxide capture process by potassium carbonate solution. *Energy*, **254**, Part A, https://doi.org/10.1016/j.energy.2022.124290

CO_2 Emissions in 2022 (2023). International Energy Agency Publications. IEA France. http://www.iea.org (accessed 10 September 2023)

Cui X. and Kær S. K. (2020). A comparative study on three reactor types for methanol synthesis from syngas and CO_2. *Chemical Engineering Journal*, **393**, 124632, https://doi.org/10.1016/j.cej.2020.124632

Czech R., Zabochnicka-Świątek M. and Świątek M. K. (2020). Air pollution as a result of the development of motorization. *Global Nest Journal*, **22**(2), 220–230, https://doi.org/10.30955/gnj.003021

Foong S. Y., Liew R. K., Yang Y., Cheng Y. W., Yek P. N. Y., Wan Mahari W. A., Lee X. Y., Han C. S., Vo D.-V. N., Van Le Q., Aghbashlo M., Tabatabaei M., Sonne C., Peng W. and Lam S. S. (2020). Valorization of biomass waste to engineered activated biochar by microwave pyrolysis: progress, challenges, and future directions. *Chemical Engineering Journal*, **389**, 124401, https://doi.org/10.1016/j.cej.2020.124401

Hong C., Xing Y., Hua X., Si Y., Qiao G. and Wang Z. (2015). Dewaterability of sludge conditioned with surfactant DDBAC pretreatment by acid/alkali. *Applied Microbiology and Biotechnology*, **99**, 6103–6111, https://doi.org/10.1007/s00253-015-6451-2

Itakura T., Sasai R. and Itoh H. (2006). Resource recovery from Nd–Fe–B sintered magnet by hydrothermal treatment. *Journal of Alloys and Compounds*, **408**, 1382–1385, https://doi.org/10.1016/j.jallcom.2005.04.088

Khandelwal H., Dhar H., Thalla A. K. and Kumar S. (2019). Application of life cycle assessment in municipal solid waste management: a worldwide critical review. *Journal of Cleaner Production*, **209**, 630–654, https://doi.org/10.1016/j.jclepro.2018.10.233

Krzywonos M., Borowski P. F., Kupczyk A. and Zabochnicka-Swiatek M. (2014). Abatement of CO_2 emissions by using motor biofuels. *Chemistry & Industry*, **93**, 1124–1127, https://doi.org/dx.medra.org/10.12916/przemchem.2014.1124

Ministry of Climate and Environment, Department of Strategy and Analysis. (2022). Eighth national communication and fifth biennial report under the United Nations framework convention on climate change, Ministry of Climate and Environment, Department of Strategy and Analysis, Warsaw. (accessed September 10, 2023).

Olivares-Marín M. and Maroto-Valer M. M. (2012). Development of adsorbents for CO_2 capture from waste materials: a review. *Greenhouse Gases: Science and Technology*, **2**, 20–35, https://doi.org/10.1002/ghg.45

Shen X., Yan F., Li C., Qu F., Wang Y. and Zhang Z. (2021a). Biogas upgrading via cyclic CO_2 adsorption: application of highly regenerable PEInano-Al2O3 adsorbents with anti-urea properties. *Environmental Science & Technology*, **55**, 5236–5247, https://doi.org/10.1021/acs.est.0c07973

Shen X., Yan F., Li C., Zhang Z. and Zhang Z. (2021b). A green synthesis of PEInano-SiO2 adsorbent from coal fly ash: selective and efficient CO_2 adsorption from biogas. *Sustainable Energy Fuels*, **5**, 1014–1025, https://doi.org/10.1039/D0SE01780A

Skurygin E. F. and Poroyko T. A. (2020). Absorption of carbon dioxide with hot potassium carbonate solution: modeling and statistical analysis of known experimental data. *Chemical Engineering Transactions*, **81**, 847–852.

Teo J. Y. Q., Ong A., Tan T. T. Y., Li X., Loh X. J. and Lim J. Y. C. (2022). Materials from waste plastics for CO_2 capture and utilisation. *Green Chemistry*, **24**, 6086–6099, https://doi.org/10.1039/D2GC02306G

Wang P., Guo Y., Zhao C., Yan J. and Lu P. (2017). Biomass derived wood ash with amine modification for post-combustion CO2 capture. *Applied Energy*, **201**, 34–44, https://doi.org/10.1016/j.apenergy.2017.05.096

Zabochnicka M. (2022). Industrial wastewater as a growth medium for microalgal biomass for a sustainable circular bioeconomy. *Applied Sciences*, **12**(20), 10299, https://doi.org/10.3390/app122010299

Zabochnicka-Świątek M. (2013). Utilization of *Chlorella vulgaris* and sediments after N-NH_4 removal containing clinoptilolite for sorption of heavy metals from wastewater. *Rocznik Ochrona Środowiska*, **15**, 324–347.

Zhan L. and Xu Z. (2014). State-of-the-art of recycling e-wastes by vacuum metallurgy separation. *Environmental Science & Technology*, **48**, 14092–14102, https://doi.org/10.1021/es5030383

doi: 10.2166/9781789065138_0015

Chapter 2

CO_2 capture using electrochemically produced ferriferrohydrosol

D. Budilovskis and I. Svitojūtė*[1]

Ineco, UAB, Antakalnio g. 48A-305, LT-10304 Vilnius, Lithuania
**Corresponding author: vynas@ineco.lt; ieva@ineco.lt*

ABSTRACT

During anodic dissolution of iron at small amounts of electrolyte, the ferriferrohydrosols (FFHs) of well-developed spatial structure are formed. FFH is composed mainly of Fe(II), Fe(III), and OH^- ions. Research has been carried out with the purpose of obtaining the most suitable composition of FFH for CO_2 capture. Depending on the electrolysis conditions the composition of FFH may be varied. The ability to sorb CO_2 depends on both the content of Fe(II) in preparation and spatial structure of the composition obtained. However, a certain amount of Fe(III) is also required to stabilize the FFH suspension. The most stable suspension is obtained when it contains 80% Fe (II) of total dissolved iron. The total amount of iron also affects the stability of the suspension. At best, it does not exceed 60 g/L. Studies showed that FFH effectively sorbs CO_2 from the combustion gases. Only Fe(II) ions bind CO_2. The higher their Fe(II) content in the suspension is, the more efficiently the CO_2 is sorbed. The formed compound of CO_2 and Fe(II) is characterized by high stability in an acidic medium. Analysis of chemical composition of precipitate formed showed that it basically consists CO_2 and Fe(II).

Keywords: Ferriferrohydrosol, electrolysis, carbon dioxide, decontamination

2.1 INTRODUCTION

Iron is the first element according to the mass of the Earth's core. Besides, it is the largest microelement in living organisms. The variety of iron compounds allows them to be used in various pollution removal technologies, obtaining harmless, stable and safely buried or easily used products. Iron mining is large and relatively cheap; many iron and its compounds are formed as production waste. Iron and its compounds are cheap, available, convenient, and ecologically acceptable reagents for various methods of pollution removal (Cundy *et al.*, 2008; Gylienė *et al.*, 2006).

Depending on the composition of pollutants, various iron compounds, zero-valent (metallic) iron, Fe(II), Fe(III), and Fe(VI) compounds are used for decontamination. Metallic iron is capable of destroying organic compounds containing unsaturated bonds, including dioxins and the synthetic dyes (Colombo *et al.*, 2015; Kulkarni *et al.*, 2008). In order to avoid iron passivation and increase in reaction rates, iron nanoparticles and their composites are used (Castro *et al.*, 2018; Huanga *et al.*, 2010; Lien & Zhang, 1999). When the free access of oxygen into solutions is allowed, the oxidative destruction of compounds is possible as well. It is assumed that in this case the active oxygen radical on the iron surface is formed, which leads to the Fenton's reaction (Noradaum & Cheng, 2005).

Fe^{2+} ad Fe^{3+} give an amorphous precipitate with exceptional coagulation properties. It acts as good sorbent for organic and inorganic substances. Fe^{3+} ions also give an insoluble precipitate with some organic compounds (Hashim *et al.*, 2011). In order to improve Fe^{2+} ad Fe^{3+} properties as decontaminants, their polymeric compounds are produced and used. One of these is $[Fe_2(OH)_n(SO_4)_{(6-n)/2}]_m$ synthesized using oxidizing agents (Fan *et al.*, 2002). Despite the particularly strong oxidizing properties of Fe(VI) it is rarely used for decontamination of pollutants due to its complex synthesis.

The use of iron-based technologies for decontamination is a rapidly developing field. Company Ineco (Lithuania) synthesized and produces a polymeric compound of Fe(II) and Fe(III)–ferriferrohydrosol (FFH) characterized by high decontamination efficiency. FFH is successfully used for the decontamination of wastewaters for the removal of the pollutants of different chemical composition including organic compounds. Making strong Fe-OC bonds enables it to remove effectively such compounds as organic acids, strong ligands (EDTA, ethylenediamine), phenols, chlorinated organic compounds, and so on (Kumpiene *et al.*, 2018). The strong Fe-OC bond in organic compounds allows us to expect that it will also form with atmospheric CO_2, especially since the naturally occurring iron carbonate, siderite, is the most stable known carbonate.

In order to avoid dramatic climate change removing CO_2 from the atmosphere is an urgent task. The main source of CO_2 emissions is the burning of fossil fuels (Buekens & Huang, 1998). It should be noted that biodegradation also accounts for a significant part of CO_2 emissions (Levis & Barlaz, 2011). Many effective methods for CO_2 capture have been proposed, but the main problem remains regarding the disposal of captured CO_2. The goal of our work is to combine CO_2 with iron into a compound that could be stable in the natural environment.

2.2 METHODS

2.2.1 FFH production ad characterization

FFH is produced electrochemically by anodic dissolution of metallic iron at the conditions of shortage of anions. FFH was produced in a purpose-built automatically controlled electrolyzer equipped with current reversal and educator mixing. Carbon steel sheets were used as electrodes.

Operating parameters of the electrolyzer were as follows: working volume 500 L, total current 500 A, voltage 2.5–3.5 V. During the research, the material concentrations, electrolysis reverse times, and current densities were changed.

The infrared spectra of the FFH precipitate were recorded in KBr pellets on a Fourier transformation infrared spectrometer (Hartman & Braun, Canada) with 2 cm^{-1} scale resolutions. The spectra were recorded in the wave number region between 4000 and 500 cm^{-1}.

2.2.2 CO_2 capture

Tests were conducted in air-insulated vessels by passing pure CO_2 through the FFH suspension used for effluent decontamination. It contained (g/L) Fe (total)–30, Fe(II)–24, NaCl–3, pH ~8.

FFH suspensions and those containing CO_2 were filtered and dried. The formed precipitates were characterized by pH-metric titration and SEM analysis.

2.3 RESULTS AND DISCUSSION

During the electrochemical process, the anode melts and the cathodic reduction process takes place on the cathode. On it, free iron ions with a very low concentration (about 20 mg/L) are reduced. Hydrogen is released mainly on the cathode. In the area near the anode, iron is dissolved, which immediately turns into hydrated iron compounds. During work, the passivation of the anode begins in the inter-electrode space (about 12 mm distance between the electrodes). When the poles are reversed, the iron sheets start working in the opposite direction and the anode is activated. FFH forms as an aqueous suspension at pH ~7–8. It contains up to 50 g/L of Fe(II) and Fe(III) and metallic iron particles. Depending on the electrolysis conditions, the amount of Fe (III) can vary from 2 to 10 wt% and that of metallic iron from 0 to 5 wt%. FFH preparation is dominated by Fe(II), which is connected to the bulk framework structure by -OH bonds, which ensures a high stability of the suspension before and after the collection of pollutants. The particle size and chemical composition of FFH can be controlled by electrolysis conditions and specific additives.

The essential difference between the use of FFH in liquid and gas cleaning is that the contact time with gas is significantly shorter. During that time, the gas must react with the FFH particles as fully as possible. One of the possible ways to achieve this is to use the smallest possible particles. The implementations of educator mixing in the FFH generation system allowed a remarkable increase in the dispersity of FFH suspensions. The obtained FFH was clearly a colloidal solution instead of suspensions. This allows the use of FFH in spry mode. In all cases the metal yield was 90–92%.

During the tests, it was found that after using educator mixing, it is possible to reduce the amount of the conductive component to 1.0–2.0 g/L, without worsening the yield parameters (typical amount of conductive component 7.0–8.0 g/L).

However, electrolysis conditions and applied additives have the greatest influence on the composition of FFH (Table 2.1). It is known that alcohols usually block cathodic and anodic processes. This blocking causes uneven dissolution of the anode and its mechanical breakdown. In this way, metallic

Table 2.1 Dependence of FFH composition on the conditions of electrolysis.

No.	Conditions of Electrolysis			FFH Compositions, g/L
	Current Density, A/dm^2	Frequency of Current Reversal, min	Additive, g/L	
1	0,5	60,0	–	Fe(II) – 22;Fe(III) – 2; Fe_{met} – 0,2; NaCl – 3
2	0,5	10,0	–	Fe(II) – 30;Fe(III) – 2; Fe_{met} – 0,4; NaCl – 3
3	10,0	60,0	–	Fe(II) – 28;Fe(III) – 2; Fe_{met} – 0,2; NaCl – 3
4	10,0	10,0	–	Fe(II) – 18;Fe(III) – 2; Fe_{met} – 0,4; NaCl – 3
5	0,5	60,0	–	Fe(II) – 30;Fe(III) – 2; Fe_{met} – 0,2; Na_2SO_4 – 3
6	0,5	10,0	–	Fe(II) – 30;Fe(III) – 2; Fe_{met} – 0,4; Na_2SO_4 – 3
7	10,0	60,0	–	Fe(II) – 25;Fe(III) – 2; Fe_{met} – 0,4; Na_2SO_4 – 3
8	10,0	10,0	–	Fe(II) – 34;Fe(III) – 2; Fe_{met} – 0,6; Na_2SO_4 – 3
9	0,5	60,0	Ethanol –1	Fe(II) – 33;Fe(III) – 2; Fe_{met} – 1,2; Na_2SO_4 – 3
10	0,5	10,0	Ethanol –2	Fe(II) – 19;Fe(III) – 2; Fe_{met} – 2; NaCl – 3
11	10,0	60,0	Glycerol –1	Fe(II) – 20;Fe(III) – 2; Fe_{met} – 2; Na_2SO_4 – 3
12	10,0	10,0	Ethyleglycol –1	Fe(II) – 20;Fe(III) – 2; Fe_{met} – 1,8; Na_2SO_4 – 3

Electrodes—carbon steel.

iron appears in the composition of FFH. The presence of metallic iron is important in the practical use of FFH for the treatment of flue gases when they contain other hazardous substances, for example dioxins.

FFH suspensions are not stable, they change with time. Their sorption properties also weaken. FTIR spectra (Figure 2.1) show that during long-term storage of FFH suspensions, there are changes in bonds between elements. The absorption bands at 742 and 1019 cm^{-1} in spectra of freshly formed precipitate (Curve 1) assigned to O–H bending vibrations in γ-FeOOH, whereas hydrated iron oxides dominated in aged precipitate (Curve 2). Characteristic absorption at 795 ir 889 cm^{-1} (O–H bending vibrations of structural hydroxyls) and below 700 cm^{-1} (Fe–O vibrations in iron oxides) are observed.

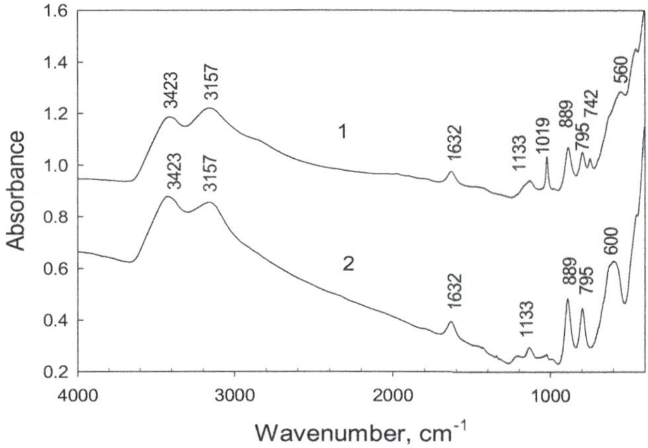

Figure 2.1 FTIR spectra of (1) freshly formed FFH precipitate and (2) after 6 months.

Table 2.2 EDS analysis of FFH precipitate before and after capture of CO_2 dried at 100°C.

Precipitate	Chemical Composition in Weight %									
	C	O	Fe	Na	Al	Si	S	Cl	Ca	Mn
FFH	2.77	37.3	58.2	0.14	0.04	0.1	0.05	0.75	0.01	0.22
FFH after CO_2 sorption	17.4	32.2	47.4	0.1	0.08	0.1	0.13	0.8	0.13	0.47

Energy-dispersive spectroscopy (EDS) investigations showed the rather big amounts of CO_2 captured by FFH, that is, ~1 mol CO_2 to 1 mol Fe (Table 2.2).

Remarkably, this newly arisen product is distinguished by high stability in acidic solutions. pH-metric titrations curves indicate that aqueous suspension of FFH (Figure 2.2, Curve 1) is sensitive to acid concentration. Likely, in this case the plateau at pH 2.2 indicates the dissolution of metallic Fe.

The plateau at pH 2.6 in the case of FFH after CO_2 capture (Figure 2.1, Curve 2) could be attributed to the interaction of carbon dioxide with acid. Likely, the newly formed iron-carbon dioxide compound is similar to natural mineral siderite (Fe(II)-carbonate), which is also distinguished by high stability in acidic solutions. The decomposition of alkaline carbonates starts at values of pH ~5. The high stability of the carbon dioxide containing compound in acidic solutions allows them to be maintained in the environment. Usually, the acidity of groundwater is less, pH is higher than 3.5. Besides, iron compounds are widely spread in the environment, biocompatible, and should not cause contamination of the surrounding areas.

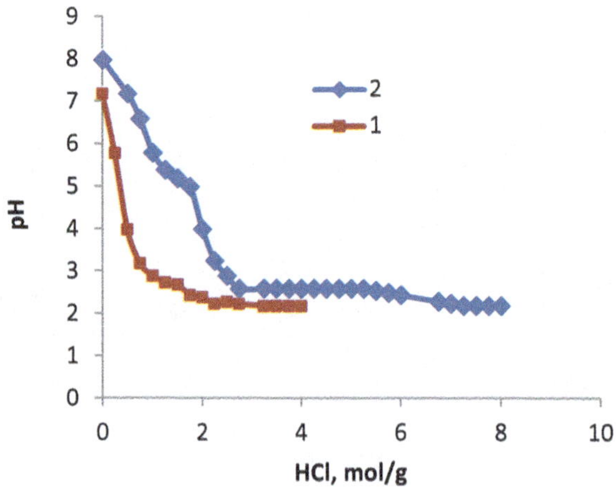

Figure 2.2 pH-metric titration of (1) FFH precipitates and (2) after CO_2 sorption.

2.4 CONCLUSIONS

Appropriate electrolyzer design, successfully selected electrolyte composition, and electrolysis conditions enable the use of ferriferrohydrosol for CO_2 capture both in suspensions and aerosols.

REFERENCES

Buekens A. and Huang H. (1998). Comparative evaluation of techniques for controlling the formation and emission of chlorinated dioxins/furans in municipal waste incineration. Review. *Journal of Hazardous Materials*, **62**, 1–33, https://doi.org/10.1016/S0304-3894(98)00153-8

Castro L., Blázquez M. L., González F., Muñoz J. A. and Ballester A. (2018). Heavy metal adsorption using biogenic iron compounds. *Hydrometallurgy*, **179**, 44–51, https://doi.org/10.1016/j.hydromet.2018.05.029

Colombo A., Dragonetti C., Magni M. and Roberto D. (2015). Degradation of toxic halogenated organic compounds by iron-containing mono-, bi- and tri-metallic particles in water. Review. *Inorganica Chimica Acta*, **431**, 48–60, https://doi.org/10.1016/j.ica.2014.12.015

Cundy A. B., Hopkinson L. and Whitby R. L. D. (2008). Use of iron-based technologies in contaminated land and groundwater remediation: a review. *Science of the Total Environment*, **400**, 42–51, https://doi.org/10.1016/j.scitotenv.2008.07.002

Fan M., Asce M., Brown R. C., Wheelock T. D. and Laabs F. C. (2002). Synthesis, characterization, and coagulation of polymeric ferric sulfate. *Journal of Environmental Engineering*, **128**, 483–490, https://doi.org/10.1061/(ASCE)0733-9372(2002)128:6(483)

Gylienė O., Šalkauskas M. and Binkienė R. (2006). Eisen und seine Verbindungen bei der Abwasserbehandlung. Jahrbuch Oberflachentechnik, **62**, 333–343.

Hashim M. A., Mukhopadhyay S., Sahu J. N. and Sengupta B. (2011). Remediation technologies for heavy metal contaminated groundwater. Review. *Journal of Environmental Management*, **92**, 2355–2388, https://doi.org/10.1016/j.jenvman.2011.06.009

Huanga H.-Y., Shiehb Y.-T., Shiha C.-M. and Twua Y.-K. (2010). Magnetic chitosan/iron (II, III) oxide nanoparticles prepared by spray-drying. *Carbohydrate Polymers*, **81**, 906–910, https://doi.org/10.1016/j.carbpol.2010.04.003

Kulkarni P. S., Crespo J. G. and Afonso C. A. M. (2008) Dioxins sources and current remediation technologies–a review. *Environment International*, **34**, 139–153, https://doi.org/10.1016/j.envint.2007.07.009

Kumpiene J., Ragnvaldsson D., Lövgren L., Tesfalidet S., Gustavsson B., Lättström A., Leffler P. and Maurice C. (2018). Impact of water saturation level on arsenic and metal mobility in the Fe-amended soil. *Chemosphere*, **74**, 206–215, https://doi.org/10.1016/j.chemosphere.2008.09.068

Levis J. W. and Barlaz M. A. (2011). Is biodegradability a desirable attribute for discarded solid waste? Perspectives from a National Landfill Greenhouse gas inventory model. *Environmental Science & Technology*, **45**, 5470–5476, https://doi.org/10.1021/es200721s

Lien H. and Zhang W. (1999). Transformation of chlorinated methanes by nanoscale iron particle. *Journal of Environmental Engineering*, **125**, 1042–1047, https://doi.org/10.1061/(ASCE)0733-9372(1999)125:11(1042)

Noradaum C. E. and Cheng I. F. (2005). EDTA degradation induced by oxygen activation in a zerovalent iron/air/water system. *Environmental Science & Technology*, **39**(18), 7158–7163, https://doi.org/10.1021/es050137v

doi: 10.2166/9781789065138_0023

Chapter 3

CO_2 capture by ferriferrohydrosol

I. Bychko[1], A. Trypolskyi[1], V. Kovbasiuk[1], A. Michuda[1], O. I. Ivanenko[2], V. Ulevičius[3], M. Zabochnicka[4] and P. Strizhak[1]**

[1]L.V. Pisarzhevskii Institute of Physical Chemistry of the National Academy of Sciences of Ukraine, Nauky Ave. 31, Kyiv, Ukraine
[2]Department of Ecology and Technology of Plant Polymers, National Technical University of Ukraine «Igor Sikorsky Kyiv Polytechnic Institute», Peremohy Ave. 37, Kyiv, Ukraine 03056
[3]State Research Institute Center for Physical Sciences and Technology, Savanorių ave. 231, Vilnius LT-02300, Lithuania
[4]Czestochowa University of Technology, Faculty of Infrastructure and Environment, Dąbrowskiego 69, Częstochowa 42-201, Poland
**Corresponding authors: igorbychko@ukr.net; psrtizhak@hotmail.com*

ABSTRACT

Commercial product ferriferrohydrosol (FFH), which is a suspension of iron oxides/hydroxides, is found applicable for the purification of wastewater to absorb CO_2 from concentrated exhaust gases. DLS, FTIR, and Raman-spectroscopy studies show that FFH consists mainly of iron-containing nanoparticles which are a mixture of Fe^{2+} and Fe^{3+} oxides and hydroxides. These nanoparticles form agglomerates with a size of several micrometers and develop a surface in the range of 200–400 m^2/g. Maghemite, hematite, and magnetite with significant content in their hydrated forms are the main crystalline phases of these particles. FFH is found to absorb CO_2 up to 0.26 $g(CO_2)/g(Fe)$ at room temperature.

Keywords: carbon dioxide, ferriferous hydrosol, hydrogenation, purification, waste gases

3.1 INTRODUCTION

The demand for strategies to reduce global atmospheric concentrations of greenhouse gases is considered one of the main tasks of the 21st century since industries release huge amounts of contaminants such as wastes, wastewaters and carbon dioxide (Bargieł & Zabochnicka-Świątek, 2018; Stańczyk-Mazanek *et al.*, 2012; Zabochnicka-Świątek, 2013; Zabochnicka, 2022). The capture and sequestration of carbon dioxide, the predominant greenhouse gas, is a central strategy of these initiatives. Carbon capture and storage (CCS) schemes embody a group of technologies for the capture of CO_2 from power plants, followed by compression, transport, and permanent storage (Raza *et al.*, 2018; Czech *et al.*, 2020; Krzywonos *et al.*, 2014). The main technologies for CO_2 capture are based on the processes of absorption–desorption in absorbing solutions, membrane separation, adsorption, and mineralization (Brunetti *et al.*, 2010; Mofarahi & Gholipour, 2014; Zhang & DePaolo, 2017). Technology for the use of absorbent solutions, in particular amines, is common in the industry, while other technologies are in the concept or pilot stages (Darunte *et al.*, 2016). Carbon capture and utilization (CCU) is an alternative approach to reducing CO_2 emissions (Ghiat & Al-Ansari, 2021). CCU can be realized by chemical fixation through the conversion of CO_2 into fuels, commodity chemicals, construction materials, or mineral carbonates represents another promising alternative for CO_2 capture. Particularly, CO_2 may be considered as feed to produce fuels such as methanol, formic acid, dimethyl carbonate, methyl formate, and higher hydrocarbons, as well as polymeric materials and pharmaceutical chemicals (Anwar *et al.*, 2020).

The carbonate cycle is a novel concept for CCU technology that is based on the adsorption of CO_2 by metal oxides with further transportation and degassing with the regeneration of initial oxide (Lux *et al.*, 2018). For the metal carbonate decarboxylation, the reaction conditions, especially the nature of the gas atmosphere, play a crucial role in the course of the reaction. If carried out in a reducing atmosphere of hydrogen, the decomposing of carbonate combines with chemical transformations of CO_2 into valuable products. The main advantage of the carbonate method is the possibility of the transformation of carbon dioxide into hydrocarbons using hydrogen (Bychko *et al.*, 2021). Using 'green' hydrogen in the hydrogenation of carbonates predicts a closed carbon cycle. Iron oxides are a leading candidate for this technology due to the high dissemination and low cost of iron. Iron is an industrial catalyst of Fisher–Tropsh synthesis, a process of obtaining synthetic fuels from synthesis gas (Frątczak *et al.*, 2021). It could be expected, that selection of experimental conditions allows obtaining not only methane and carbon monoxide, as reaction products, but also higher hydrocarbons. Iron carbonate also can be utilized as an inorganic pigment that provides a brown and red-brownish coloring.

As reported in the literature, only Fe^{2+}, mainly in the form of FeO, can capture CO_2, while Fe^{3+} does not react with CO_2 due to its inability to form carbonates. However, FeO is a quasi-stable oxide that disproportionates producing Fe and Fe_3O_4 (Mendoza *et al.*, 2019a). Therefore, iron in the form of Fe_3O_4, or a mixture of iron oxides/hydroxides is suitable for large-scale adsorption of CO_2 (Mendoza *et al.*, 2019b). Adsorption capacity and kinetics can be increased by the use of a water suspension. The commercial products of suspension of iron oxides/hydroxides named ferriferrohydrosol (FFH) which is used for purification of wastewater, is also predicted to be a perspective absorbent for CO_2. FFH is a colloidal suspension of two- and three-valence iron hydrated compounds, used as a reagent for wastewater pollutants binding. The method of manufacturing FFH is simple and based on the electrolysis of iron and steel stamping waste. Predicted, that electrolysis of steel leads to a formation of iron-containing oxide–hydroxide nanoparticles that shows a tendency of coagulation into coarse bunches, and the formation of agglomerates with a size of several micrometers. Formed FFH characterizes by enhanced sorption capacity, particularly due to the developed specific surface of iron-containing particles, that can achieve 400 m^2.

The present work is dedicated to the assessment of the possibility of CO_2 capturing using the FFH suspension as a perspective product for cleaning concentrated exhaust gases produced by fossil fuel-based enterprises.

3.2 METHODS

Samples of FHH were provided by the manufacturer, the INECO company (Innovation Ecology), Vilnius, Republic of Lithuania (Figure 3.1). It was provided three samples that contain 40 g/l of FFH with 4 g/l of NaCl (FFH-NaCl), 30 g/l of FFH with 4 g/l of Na_2SO_4 (FFH-Na_2SO_4), and 40 g/l of FFH with 4 g/l of Na_2SO_4 and 2 g/l of a compound (FFH-Na_2SO_4+). Also, commercial magnetite Fe_3O_4 325 mesh ($\approx$45 μm) from 'Thermo scientific' was used as a sample for comparison. The structure of the samples was characterized by Raman spectroscopy. Raman spectra were performed with HORIBA Jobin-Yvon T64000 spectrometer using exciting Ar–Kr laser radiation at a wavelength of 488 nm and a power of 100 mW. Fourier transform infrared (FTIR) spectra were obtained using a Spectrum-One spectrometer (PerkinElmer). The dynamic light scattering (DLS) measurements were made with a Malvern Zetasizer Nano S, using a 633 nm HeNe laser at 4 mW. Aqueous suspensions of initial FFH, and FFH after CO_2 adsorption were tested in quartz cuvettes having a 10 mm path length. The analysis was operated in backscatter mode at an angle of 173°. Samples were equilibrated at 25°C for 30 min before measurement. The concentration of FFH in water was 1 mg/ml. The capacity of FFH for CO_2 adsorption was determined using a volumetric method. The flask with a sample was connected to a cuvette filled with dibutylpthalate. The duration of each adsorption experiment was at least 30 min.

Figure 3.1 Sample of FFH.

3.3 RESULTS AND DISCUSSION

Figure 3.2 presents FTIR and Raman-spectra of initial FFH samples, commercial Fe_3O_4, and corresponding samples after CO_2 adsorption. A comparison of spectra of initial FFH shows that spectra of FFH-Na_2SO_4+ and FFH-Na_2SO_4 are very similar, while the spectrum of FFH-NaCl shows a noticeable difference. Bands at 340 cm^{-1}, 495 cm^{-1}, and a broadband at 685 cm^{-1} in combination with a band at 1395 cm^{-1} show that the main phase of FFH-Na_2SO_4+ and FFH-Na_2SO_4 is maghemite—γ-Fe_2O_3 (de Faria *et al.*, 1997). The spectra of FFH-NaCl present the same bands combined with two intense bands at 220 and 280 cm^{-1} which corresponds to hematite—α-Fe_2O_3 (Shim & Duffy, 2002). Raman spectra of magnetite contain a weak band at 540 cm^{-1} and a strong band at 662 cm^{-1}, which are typical bands of magnetite (Chicot *et al.*, 2009).

A comparison of Raman-spectra of initial FFH samples and FFH after CO_2 adsorption shows only a slight decrease of band intensity. The main Raman bands of siderite appear at 190 cm^{-1}, 290 cm^{-1}, and the most intense band at 1100 cm^{-1} (Langille & O'Shea, 1977). Therefore, CO_2 adsorption does not lead to the appearance of detectable bands of siderite, which can be caused by a low concentration of formed siderite. Raman spectra of magnetite after CO_2 adsorption are almost identical to the initial Fe_3O_4.

Bands at 570–580 cm^{-1} in the FTIR spectra correspond to a Fe–O stretching mode of the tetrahedral and octahedral sites of Fe_3O_4 in Fe_3O_4, and FFH samples. Weak band at 800 cm^{-1} can be attributed to a -OH stretching vibration

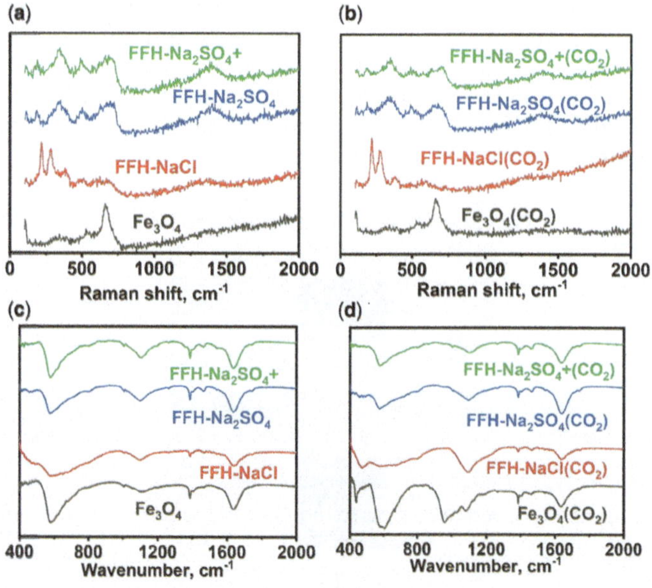

Figure 3.2 Raman (a) and FTIR (b) spectra of initial FFH samples, commercial Fe_3O_4, and Raman (c) and FTIR (d) spectra of corresponding samples after CO_2 adsorption.

of α-FeOOH (Stoia *et al.*, 2016; Xing *et al.*, 2020). Bands with low intensity near 1000 cm^{-1} are the bending vibration of –OH modes in γ-FeOOH and a broad peak at 1100 cm^{-1} is the bending vibration of OH modes in δ-FeOOH. Peaks at 1300–1500 cm^{-1} could be related to the stretching vibration band of the CO_3 group that mainly arises from the contamination of solutions by atmospheric carbon dioxide (Kumar *et al.*, 2015). A pronounced band at ~1620–1630 cm^{-1} is related to the H–O–H bending of water.

Adsorption of CO_2 leads to the appearance of the band at 470 cm^{-1} in FFH-NaCl(CO_2) and FFH-Na_2SO_4(CO_2) as well as a new band at 440 cm^{-1} in Fe_3O_4-CO_2, which corresponds to metal-oxygen vibration modes of maghemite (Sankadiya *et al.*, 2016). The spectra of Fe_3O_4-CO_2 contain a doubled band at 520–750 cm^{-1} which is a result of the combination of Fe–O stretching mode of Fe_3O_4 and γ-Fe_2O_3. The spectra of Fe_3O_4(CO_2) contain two intense bands at 900–1100 cm^{-1} that era attributed to surface OH groups in magnetite (960 cm^{-1}) and maghemite (1080 cm^{-1}). There is a significant increase in the intensity of the band at 1100 cm^{-1} for the FFH-NaCl(CO_2) sample in comparison with the initial FFH-NaCl. This result indicates an increase in the content of the γ-FeOOH phase due to CO_2 adsorption. Adsorption of CO_2 does not lead to significant changes in the FTIR spectra of FFH-Na_2SO_4(CO_2) and of FFH-Na_2SO_4+ (CO_2) compared with initial samples.

Table 3.1 presents the average particles size of initial FFH samples and FFH samples after CO_2 adsorption determined by DLS. The obtained results show that FFH contains particles of a size of several micrometers. The smaller particles of 2.5 μm contain FFH-NaCl, while FFH-Na_2SO_4+ contains the larger particles of 6.5 μm. Adsorption of CO_2 decreases the average particle size for all samples to a size close to 2.7 μm. This result indicates that the adsorption of CO_2 leads to breaking the bonds in iron-containing particles, probably hydrogen bonds that result in the grinding of the particles agglomerates to smaller initial particles.

Therefore, FFH contains a suspension of iron-containing particles with a size of several micrometers which are a mix of oxides and hydroxides of Fe^{2+} and Fe^{3+}. The main crystalline phases of these particles are maghemite, hematite, and magnetite with a significant content of their hydrated forms. It can be speculated, that the main amount of CO_2 adsorbs by Fe in amorphous forms, which are the major phase of Fe in FFH. Adsorption of CO_2 has a limited effect

Table 3.1 Average particles size of initial FFH samples and FFH after adsorption of CO_2.

No.	Sample	Average Particles Size, μm
1	FFH-Na_2SO_4+	6.5 ± 0.2
2	FFH-Na_2SO_4	3.4 ± 0.1
3	FFH-NaCl	4.7 ± 0.1
4	FFH-Na_2SO_4 + (CO_2)	2.8 ± 0.1
5	FFH-Na_2SO_4(CO_2)	2.7 ± 0.1
6	FFH-NaCl(CO_2)	2.6 ± 0.1

Table 3.2 Density of FFH samples, the amount of adsorbed CO_2 by FFH samples at 20°C, and the amount of adsorbed CO_2 relative to the theoretic value obtained from the proportion of (1 mole CO_2)/(1 mole FeO).

Sample	Density, g/cm³	Liter(CO_2)/ Liter(FFH)	g(CO_2)/ kg(FFH)	g(CO_2)/ g(Fe)	% From Theory
FFH-Na_2SO_4+	1.06	3.1	5.8	0.15	8.5
FFH-Na_2SO_4	1.04	3.9	7.4	0.26	14.5
FFH-NaCl	1.05	4.3	8	0.21	12
*Fe_3O_4	-	<0.01	<0.1	<0.001	<0.001

*Experiment of CO_2 adsorption by Fe_3O_4 was provided using a suspension of 40 g Fe_3O_4 and 4 g NaCl in 1 liter of water.

on the transformation of crystalline phases, which consists in the transition of the magnetite phase to maghemite. While the amorphous phase of particles adsorbs CO_2 by surface OH-groups and oxides forming nanostructures containing carbonates and hydroxycarbonates that aggregate into particles with sizes near 2.5–3 μm.

Table 3.2 presents experimental results of the adsorption of CO_2 by FFH samples. At 20°C FFH adsorbs 3–4.5 liters or 6–8.5 g of CO_2. The ultimate amount of adsorbed CO_2 by iron oxide/hydroxide systems can be calculated according to the ratio, where 1 mole of Fe adsorbs 1 mole of CO_2. This proportion follows from the equation $FeO + CO_2 = FeCO_3$. Therefore, 1 liter of FFH, which contains 40 g of Fe, can adsorb 31.4 g or 16 liters of CO_2. Adsorption of CO_2 by water is negligible due to the influence of electrolyte and solid particles, which dramatically reduce the adsorption capacity of the water. Therefore, according to this approach, the adsorption capacity of FFH samples is 8.5–14.5% of the theoretical value. This indicates that only a limited amount of iron compounds adsorbs CO_2. The adsorption capacity determined experimentally is lower than the theoretic value, which can be caused by the inaccessibility of the inner part of particles to the CO_2 at 20°C. The high content of Fe^{3+} is another reason of the depressed capability of FFH toward the adsorption of CO_2. Adsorption of CO_2 by magnetite is negligible, that is a consequence of a very low specific surface due to large particle size.

Comparison of data presented in Tables 3.1 and 3.2 shows that amount of adsorbed CO_2 increases with a decrease in particle size of FFH. This result indicates that the influence of particle size on the adsorption capacity is very significant. Therefore, a strategy for the creation of a highly effective FFH is not only to obtain a FFH with a high content of Fe^{2+} but also with a small size of iron-containing particles. Also, it can be concluded that the main role of electrolytes is the determination of particles size of FFH, and the influence of various electrolytes on differences in phase composition, the ratio of Fe^{2+}/Fe^{3+}, and, as a consequence, the adsorption capacity is negligible.

The study of model systems of iron oxides shows that at room temperature is mainly observed physical adsorption of CO_2 on all types of iron oxides. The degree of surface hydroxylation plays a decisive role in the adsorption of CO_2

on the surface of iron oxides, which leads to the formation of adsorbed species identified as surface bicarbonates. Adsorbed CO_2 over a long period of time can be transformed into a chemisorbed form with the formation of carbonates and bicarbonates. However, that process occurs only in the presence of Fe^{2+} (Li & Paier, 2020; Mirabella *et al.*, 2018; Xu *et al.*, 2020). Accordingly, a developed surface of iron oxide is required for an increase in the absorption capacity. The maximum amount of 3.01×10^{-3} gCO_2/g of absorbed CO_2 was observed for Fe_2O_3. The adsorption of CO_2 by a biochar/Fe oxyhydroxide sorbent of 0.16 gCO_2/g at 25°C was achieved (Xu *et al.*, 2020).

Following this approach, obtained values of adsorbed CO_2 by FFH allow the determination of the specific surface area (SSA) of iron particles which can be calculated using equation (3.1):

$$S = q * \sigma_0 * N_A \tag{3.1}$$

where q is the adsorption capacity of CO_2 (mol/g), σ_0 is the area claimed by an adsorbed molecule of CO_2 (0.109×10^{-18} m^2), and N_A is Avogadro's number (6.022×10^{23} mol^{-1}). The obtained result of the SSA of iron in FFH gives 220–390 m^2/g. Comparing the values of SSA and particles size obtained by DLS confirms that the iron-containing oxide particles in the samples are porous agglomerates of smaller, mostly amorphous nanoparticles with a highly developed surface.

3.4 CONCLUSIONS

The possibility of applying FFH for the absorption of CO_2 from concentrated exhaust gases was demonstrated. It was shown, that the structure of FFH is a suspension of iron-containing nanoparticles which are a mix of oxides and hydroxides of Fe^{2+} and Fe^{3+}. These nanoparticles are agglomerated in larger particles with a size of several micrometers. The main crystalline phases of these particles are maghemite, hematite, and magnetite with significant content in their hydrated forms. FFH samples have a developed surface in the range of 200–400 m^2/g. The absorption capacity for CO_2 of 0.26 g(CO_2)/g(Fe) at room temperature was shown. A high absorption capacity of FFH can be associated with a highly developed surface area of the samples and a high degree of surface hydroxylation. Therefore, the use of FFH to absorption of CO_2 from concentrated exhaust gases has significant advantages over solid sorbents.

ACKNOWLEDGMENTS

This work was supported in the framework of the EUREKA NETWORK project "Capture of carbon dioxide by innovative sorbent InnoCO_2sorbent (EUREKA/InnoCO_2Sorbent/2/2021), financed by the Polish National Centre for Research and Development. The authors are also grateful to the EUREKA support (E! 13401 FFH-CO_2). The Ukrainian authors are thankful to the Armed Forces of Ukraine for serving our country and protecting our freedoms.

REFERENCES

Anwar M. N., Fayyaz A., Sohail N. F., Khokhar M. F., Baqar M., Yasar A., Rasool K., Nazir A., Raja M. U. F., Rehan M., Aghbashlo M., Tabatabaei M. and Nizami A. S. (2020). CO_2 utilization: turning greenhouse gas into fuels and valuable products. *Journal of Environmental Management*, **260**, 110059, https://doi.org/10.1016/j.jenvman.2019.110059

Bargieł P. and Zabochnicka-Świątek M. (2018). Technologies of coke wastewater treatment in the frame of legislation in force. *Ochrona Środowiska i Zasobów Naturalnych*, **29**(1), 11–15, https://doi.org/10.2478/oszn-2018-0003

Brunetti A., Scura F., Barbieri G. and Drioli E. (2010). Membrane technologies for CO_2 separation. *Journal of Membrane Science*, **359**(1–2), 115–125, https://doi.org/10.1016/j.memsci.2009.11.040

Bychko I. B., Kovbasiuk V. I., Trypolsky A. I., Ivanchuk V. Y. and Strizhak P. E. (2021). Low-temperature hydrogenation of iron carbonate followed by production of C4–C6 hydrocarbons. *Theoretical and Experimental Chemistry*, **57**, 351–357, https://doi.org/10.1007/s11237-021-09704-3

Chicot D., Roudet F., Lepingle V. and Louis G. (2009). Strain gradient plasticity to study hardness behavior of magnetite (Fe_3O_4) under multicyclic indentation. *Journal of Materials Research*, **24**(3), 749–759, https://doi.org/10.1557/jmr.2009.0098

Czech R., Zabochnicka-Świątek M., and Świątek M.K. (2020). Air pollution as a result of the development of motorization. *Global Nest Journal*, **22**(2), 220–230, https://doi.org/10.30955/gnj.003021

Darunte L. A., Walton K. S., Sholl D. S. and Jones C. W. (2016). CO_2 capture via adsorption in amine-functionalized sorbents. *Current Opinion in Chemical Engineering*, **12**, 82–90, https://doi.org/10.1016/j.coche.2016.03.002

de Faria D. L. A., Venâncio Silva S. and de Oliveira M. T. (1997). Raman microspectroscopy of some iron oxides and oxyhydroxides. *Journal of Raman Spectroscopy*, **28**, 873–878, https://doi.org/10.1002/(SICI)1097-4555(199711)28:11<873::AID-JRS177>3.0.CO;2-B

Frątczak J., de Paz Carmona H., Tišler Z., Hidalgo Herrador J. M. and Gholami Z. (2021). Hydrocracking of heavy Fischer–Tropsch wax distillation residues and its blends with vacuum gas oil using phonolite-based catalysts. *Molecules*, **26**, 7172, https://doi.org/10.3390/molecules26237172

Ghiat I. and Al-Ansari T. (2021). A review of carbon capture and utilisation as a CO_2 abatement opportunity within the EWF nexus. *Journal of CO_2 Utilization*, **45**, 101432, https://doi.org/10.1016/j.jcou.2020.101432

Krzywonos M., Borowski P.F., Kupczyk A. and Zabochnicka-Świątek M. (2014). Abatement of CO2 emissions by using motor biofuels. *Przemysl Chemiczny*, **83**(7), 1124–1127, https://doi: dx.medra.org/10.12916/przemchem.2014.1124

Kumar R., Sakthivel R., Behura R., Mishra B. K. and Das D. (2015). Synthesis of magnetite nanoparticles from mineral waste. *Journal of Alloys and Compounds*, **645**, 398–404, https://doi.org/10.1016/j.jallcom.2015.05.089

Langille D. B. and O'Shea D. C. (1977). Raman Spectroscopy studies of antiferromagnetic $FeCO_3$ and related carbonates. *Journal of Physics and Chemistry of Solids*, **38**, 1161–1171, https://doi.org/10.1016/0022-3697(77)90044-0

Li X. and Paier J. (2020). Vibrational properties of CO_2 adsorbed on the Fe_3O_4 (111) surface: insights gained from DFT. *The Journal of Chemical Physics*, **152**, 104702, https://doi.org/10.1063/1.5136323

Lux S., Baldauf-Sommerbauer G. and Siebenhofer M. (2018). Hydrogenation of inorganic metal carbonates: a review on its potential for carbon dioxide utilization and emission reduction. *ChemSusChem*, **11**(19), 3357–3375, https://doi.org/10.1002/cssc.201801356

Mendoza E. Y. M., Santos A. S., López E. V., Drozd V., Durygin A., Chen J. and Saxena S. K. (2019a). Iron oxides as efficient sorbents for CO_2 capture. *Journal of Materials Research and Technology*, **8**(3), 2944–2956, https://doi.org/10.1016/j.jmrt.2019.05.002

Mendoza E. Y. M., Sarmiento Santos A., Vera López E., Drozd V., Durygin A., Chen J. and Saxena S. K. (2019b). Siderite formation by mechanochemical and high pressure–high temperature processes for CO_2 capture using iron ore as the initial sorbent. *Processes*, **7**(10), 735, https://doi.org/10.3390/pr7100735

Mirabella F., Zaki E., Ivars-Barceló F., Schauermann S., Shaikhutdinov S. and Freund H.-J. (2018). CO_2 Adsorption on magnetite Fe_3O_4(111). *The Journal of Physical Chemistry C*, **122**(48), 27433–27441, https://doi.org/10.1021/acs.jpcc.8b08240

Mofarahi M. and Gholipour F. (2014). Gas adsorption separation of CO_2/CH4 system using zeolite 5A. *Microporous and Mesoporous Materials*, **200**, 1–10, https://doi.org/10.1016/j.micromeso.2014.08.022

Raza A., Gholami R., Rezaee R., Rasouli V. and Rabiei M. (2018). Significant aspects of carbon capture and storage–a review. *Petroleum*, **5**(4), 335–340, https://doi.org/10.1016/j.petlm.2018.12.007

Sankadiya S., Oswal N., Jain P. and Gupta N. (2016). Synthesis and characterization of Fe_2O_3 nanoparticles by simple precipitation method. *AIP Conference Proceedings*, **1724**, 020064, https://doi.org/10.1063/1.4945184

Shim S.-H. and Duffy T. S. (2002). Raman spectroscopy of Fe_2O_3 to 62 GPa. *American Mineralogist*, **87**, 318–326, https://doi.org/10.2138/am-2002-2-314

Stańczyk-Mazanek E., Nalewajek T. and Zabochnicka M. (2012). Drug-resistant microorganisms in soils fertilized with sewage sludge. *Archives of Environmental Protection*, **38**(1), 97–102, doi: 10.2478/v10265-012-0008-6

Stoia M., Istratie R. and Păcurariu C. (2016). Investigation of magnetite nanoparticles stability in air by thermal analysis and FTIR spectroscopy. *Journal of Thermal Analysis and Calorimetry*, **125**(3), 1185–1198, https://doi.org/10.1007/s10973-016-5393-y

Xing B., Graham N. and Yu W. (2020). Transformation of siderite to goethite by humic acid in the natural environment. *Communications Chemistry*, **3**, 38, https://doi.org/10.1038/s42004-020-0284-3

Xu X., Xu Z., Gao B., Zhao L., Zheng Y., Huang J., Tsang D. C. W., Ok Y. S. and Cao X. (2020). New insights into CO_2 sorption on biochar/Fe oxyhydroxide composites: kinetics, mechanisms, and *in situ* characterization. *Chemical Engineering Journal*, **384**, 123289, https://doi.org/10.1016/j.cej.2019.123289

Zabochnicka-Świątek M. (2013). Utilization of *Chlorella vulgaris* and sediments after N-NH_4 removal containing clinoptilolite for sorption of heavy metals from wastewater. *Rocznik Ochrona Srodowiska/Annual Set the Environment Protection*, **15**(1), 324–347

Zabochnicka M. (2022). Industrial wastewater as a growth medium for microalgal biomass for a sustainable circular bioeconomy. *Applied Sciences*, **12**(20), 10299, https://doi.org/10.3390/app122010299

Zhang S. and DePaolo D. J. (2017). Rates of CO_2 mineralization in geological carbon storage. *Accounts of Chemical Research*, **50**(9), 2075–2084, https://doi.org/10.1021/acs.accounts.7b00334

doi: 10.2166/9781789065138_0033

Chapter 4

Steam gasification of solid organic materials and conversion of obtained synthesis gas into hydrogen

O. M. Dudnyk[1,*] and I. S. Sokolovska[1,2]

[1]Thermal Energy Technology Institute of National Academy of Sciences of Ukraine, 19, Andriivska Str., Kyiv, Ukraine
[2]General Energy Institute of National Academy of Sciences of Ukraine, 172, Antonovych Str., Kyiv, Ukraine
*Corresponding author: aldudnyk2018@gmail.com

ABSTRACT

Today, the most efficient use of water for electricity generation with high efficiency is achieved at integrated gasification combined cycle (IGCC) power plants. An important feature of integrated gasification is solid and liquid fuels conversion into hydrogen-rich synthesis gas. In the case of catalytic steam conversion of CO by the water shift reaction and capture of the obtained carbon dioxide, clean hydrogen is produced at such power plants. Hydrogen thermal power plants (TPPs) are the most ecologically friendly compared to conventional TPPs.

The aim of this research was to determine the parameters of H_2 and CO_2 production as a result of solid fuel steam gasification, catalytic steam conversion of the produced synthesis gas, and CO_2 capture in the gasifier and in the low-temperature sorption reactor after the water shift reactor. Lime and calcined dolomite in the gasifier and an aqueous solution of monoethanolamine in a low-temperature CO_2 sorption reactor were used for CO_2 capture. Hydrogen up to 98% purity was obtained with the use of steam gasifier, water-shift reactor, and low-temperature CO_2 sorption reactor. 92–94% of carbon was removed in the system of hydrogen production: 37–48% – in the gasifier, 44–56% – in the low-temperature CO_2 sorption reactor.

Keywords: capture, gasification, hydrogen, synthesis gas, water, CO_2

4.1 INTRODUCTION

The reduction of carbon dioxide emissions at modern fossil fuel-fired thermal power plants (TPP) is achieved, first of all, as a result of increasing their efficiency – reducing the consumption of conventional fuel per 1 kWh of generated electricity or GJ of generated heat. Increasing the efficiency of condensing coal-fired power plants is achieved because of ensuring the quality of coal, as well as the introduction of new integration gasification combined cycle (IGCC) technologies and TPP operating at ultra-supercritical steam parameters that can reduce CO_2 emissions by 10–25%. Further reduction of CO_2 emissions at such TPP can be achieved due to the use of high-temperature fuel cell technologies. In this case, CO_2 emissions can be reduced by 30–50% (Dudnyk, 2014).

The calculation results of the influence of the composition of the combustible mass of various types of organic fuel on the yield of carbon dioxide depending on the efficiency of the use of the energy of the combustible mass of fuels in TPP are shown in the research study by Dudnyk (2014).

Table 4.1 shows the composition of the combustible mass of various types of organic fuels (from solid to gaseous).

From Table 4.1, it can be seen that the hydrogen content in organic fuels depends on the aggregate state of organic fuels and increases from 1.8 wt.% (for anthracite) up to 25.1 wt.% (for methane). As the degree of coal metamorphism increases from lignite to anthracite, the hydrogen content decreases from 5.8 to 1.8 wt.%.

Figure 4.1 shows the results of calculations of carbon dioxide output depending on the efficiency of using the energy of the combustible mass of organic fuels in TPP (Dudnyk, 2014).

After the analysis of Figure 4.1, it can be seen that the main influence on the yield of carbon dioxide per kWh of produced energy is exerted by the hydrogen content of the fuel. With an increase in the content of hydrogen in organic fuels, the yield of carbon dioxide at 1 kWh of produced energy decreases. The exception is wood. The composition of the fuel is a determining factor affecting the output of carbon dioxide gas in terms of kWh of produced energy. At the same efficiency of the use (35%), carbon dioxide emissions during the use of: anthracite by 1.1, brown coal by 1.2, fuel oil by 1.4, and methane by 2 times less than when using pure carbon.

Table 4.1 Composition of the combustible mass of various types of organic fuels (Dudnyk, 2014).

Fuel	Content of Combustible Components					Heat Value
	C	H	O	N	S	
	wt. %					MJ/kg
Carbon	100.0	–	–	–	–	33.9
Anthracite	93.5	1.8	1.5	0.8	2.4	33.7
Bituminous coal	79.0	5.2	9.9	1.5	4.4	32.7
Brown coal	61.6	5.8	25.8	0.8	6.0	24.7
Wood	50.0	6.0	43.0	1.0	–	18.5
Fuel oil	87.7	10.8	0.4	0.4	0.7	40.9
Methane	74.9	25.1	–	–	–	51.3

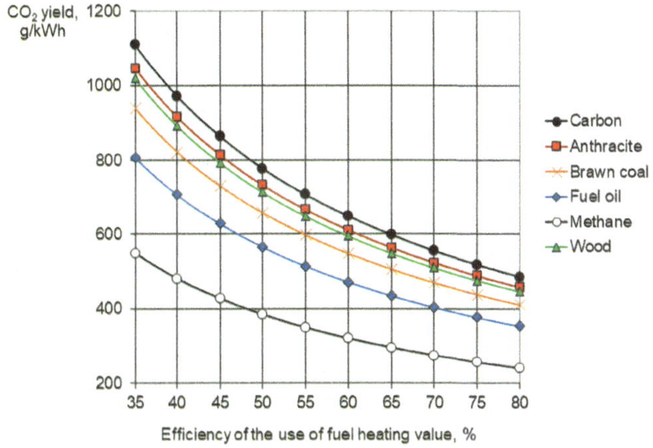

Figure 4.1 Dependence of CO_2 yield on fuel type and efficiency of the use of fuel heating value (Dudnyk, 2014).

Figure 4.2 shows the reduction in carbon dioxide emissions depending on the efficiency of energy use at TPPs (Dudnyk, 2014). The efficiency of a traditional coal-fired condensing steam turbine plant – 35% – was chosen as a starting point. When using anthracite with an efficiency of 35%, the level of carbon dioxide emissions is 1047 g/kWh, and in the case of using lignite – 940 g/kWh of generated electrical energy.

The introduction of new IGCC TPPs with the conversion of coal into hydrogen-rich synthesis gas, as well as the use of boilers with ultra-supercritical steam parameters, will ensure an increase in the electrical efficiency of power units to 40–47% and will allow reducing the level of carbon dioxide emissions by 12.5–22.5%. In this case, emissions of carbon dioxide into the environment

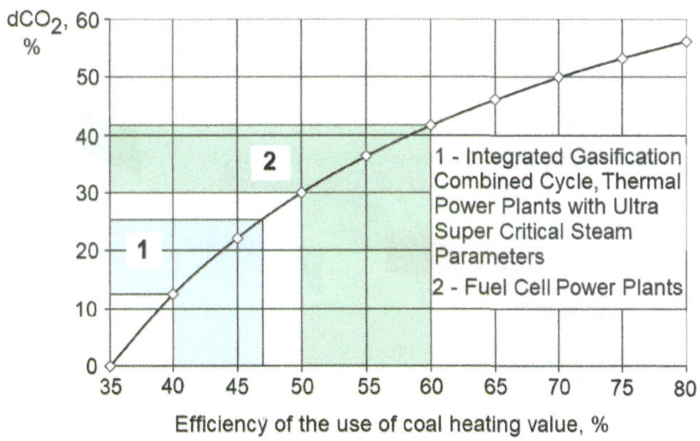

Figure 4.2 Reduction of carbon dioxide air pollution in case of the use of advanced coal energy technologies (Dudnyk, 2014).

during the use of anthracite will be 782–916 g/kWh, and brown coal will be 701–823 g/kWh of electrical energy. The use of coal-fired IGCC TPPs with fuel cells will increase the efficiency of converting coal energy into electrical energy to 50–60% and reduce CO_2 emissions into the environment by 30–50%. In this case, emissions of carbon dioxide into the environment during the use of anthracite will be 611–733 g/kWh, and brown coal will be 548–658 g/kWh of generated electrical energy.

According to the British Petroleum Review (BP Statistical Review, 2022), primary energy consumption in the Polish economy in 2021 by type of fuel was, %: oil – 31.1, natural gas – 18.9, coal – 42.3, hydroelectric power – 0.5, renewable energy sources – 7.2. Primary energy consumption in the Ukraine economy in 2021 by type of fuel (BP Statistical Review, 2022) was, %: oil – 13.8, natural gas – 28.1, coal – 28.4, nuclear energy – 23.4, hydroelectric power – 3.0, renewable energy sources – 3.3. For the data of the International Trade Administration, electricity production in 2021 was the highest in Poland's history, at 179.4 terawatt-hours (TWh). Electricity demand was also a record 180.3 TWh (Official Poland – Country Commercial Guide, 2022). Net imports of electricity were the lowest in 5 years. 99.5% of electricity demand was provided by domestic generation. The share of coal in electricity generation in 2021 increased and was over 72%. Despite high CO_2 prices, coal-fired generation was less expensive than natural gas-fired generation, resulting in a record increase in the use of coal-fired capacity and a decrease in the use of natural gas-fired capacity. For the first time in years, wholesale electricity prices in Poland were among the lowest in this part of Europe. This resulted in high exports and production. The weighted average price of CO_2 in 2021 was $62.85/t CO_2. Poland's revenue from the sale of CO_2 allowances was more than $6.5 billion in 2021.

Figure 4.3 shows the electrical efficiency of modern energy systems (Herbst, 2015). Due to the direct conversion of the chemical energy of the fuel into

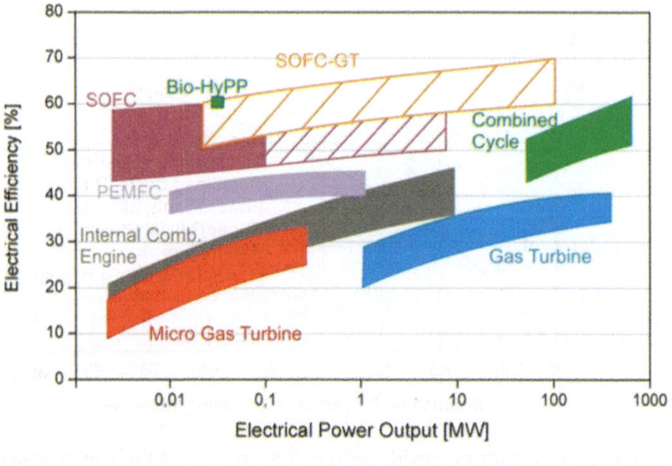

Figure 4.3 Electrical efficiency of modern power systems (Herbst, 2015).

electrical energy, SOFC power plants of relatively small electrical power (up to 100 kW) have the same and higher electrical efficiency compared to modern combined cycle TPPs (which have electric power up to 800 MW).

According to the data given in Wang & Stiegel (2016), the use of IGCC technologies for the operation of coal-fired power plants, in comparison with existing modern coal-fired power plants, will reduce emissions of sulfur oxides by 11.4 times, nitrogen oxides by 2.3 times, and dust by 100 times (Figure 4.4).

In the world at TPP for the removal of CO_2, there are three main approaches and combinations: from flue gases after convention boilers; from mixture of steam and CO_2 after OXY-Fuel boilers; from the synthesis gas, obtained after reforming of natural gas or gasification of solid and liquid fuels; from biogas after fermentations. Technologies for carbon dioxide recovery are also divided into two groups: CO_2 treatment after scalding and before scalding (Dudnik, 2012).

The processes of CO_2 separation and purification uses of adsorption with pressure change (PSI), cleaning in water scrubbers, physical adsorption in organic solvents, chemical adsorption in organic solvents, membrane separation, low temperature separation (Xiaodong, 2018).

The main industrial applications of CO_2 are for increasing the productivity of old oil and gas wells. CO_2 is used as working fluid in the modern power plants and for obtaining value-added products (CO_2 Power Solutions; Esposito, 2017; Zhu, 2019).

The basic CO_2 conversion technologies for obtaining value-added products are electrochemical, catalytic, photo-catalytic, photo-thermal catalytic, biological, copolymerization and mineral-carbonization (Zhu, 2019).

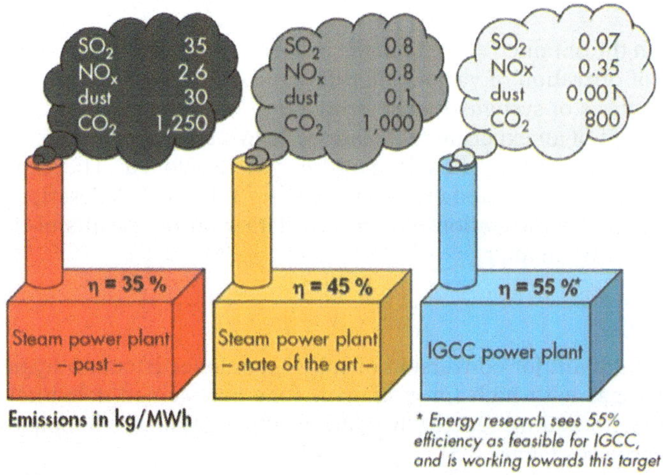

Figure 4.4 Harmful emissions of coal-fired TPP with gas cleaning systems (Wang & Stiegel, 2016).

In the world, most technologies for the production, purification, and utilization of carbon dioxide have already been tested in various industries on a commercial scale. Further introduction of these technologies in the latest energy-chemical combined complexes for the production of energy and chemical products, taking the properties of the source organic fuels into account, will significantly reduce the impact of carbon dioxide on the environment and climate change on Earth.

The analysis was carried out and proposals were prepared on the possibility of using the by-product of solid organic wastes conversion – carbon dioxide for the operation of polygeneration power plants for the production of electricity, fertilizers, plastics, third-generation biofuels, construction materials, nanocarbon, and so on (Dudnyk *et al.*, 2020).

In the world, the lowest cost of CO_2 removal in solid fuel conversion processes is achieved using IGCC TPPs. Reduced cost and improved efficiency of CO_2 sequestration are provided by removing CO_2 not from combustion products, but from solid fuel gasification products – synthesis gas. As part of the implementation of the third stage of the Osaki CoolGen project (OSG Japan), it is demonstrated at the 166 MW_e IGCC power unit the possibility of achieving electrical efficiency at the level of 47.0% with 90% capture of carbon dioxide for power units with an electrical capacity of 500 MW_e at IGCC TPP (Beginning, 2019). The high electrical efficiency of the coal-fired power unit with CO_2 capture is planned to be obtained due to the use of IGCC and SOFC technologies. In Japan, in order to cover the electrical capacity of the power units of the Fukushima NPP as part of the restoration project, the construction of two Fukushima combined cycle gasification plants with an electrical capacity of 543 MW each has begun (Wakabayashi, 2017).

The increased degree of the use of steam is ensured by two-stage gasification of solid fuels and solid organic waste (Zhovtyansky *et al.*, 2013, 2023, Dudnyk *et al.*, 2025).

Based on the obtained data about the influence of the fuel type and the energy efficiency of operation of various thermal power plants on the yield of carbon dioxide, analysis of systems for the production, purification and utilization of CO_2, a concept of an experimental plant with two-stage (high-temperature and low-temperature) carbon dioxide capture was developed. The experimental plant was created and tested using four types of solid fuel. A description of this plant, methods for estimation of obtained data, and the results of the studies are shown in this chapter.

4.2 METHODS

Experimental plant for conversion of solid organic feedstock in hydrogen and carbon acid gas with high (in gasifier by lime or calcined dolomite) and low temperature (in adding reactor by water solution of monoethanolamine) CO_2 capture was used (Figure 4.5).

The installation consists of main equipment: bunkers with coal and absorbent, gasifier, steam generator, synthesis gas cooler, tank with condensate, drier, reactor for catalytic conversion of CO, reactor for low-temperature sorption of CO_2, and

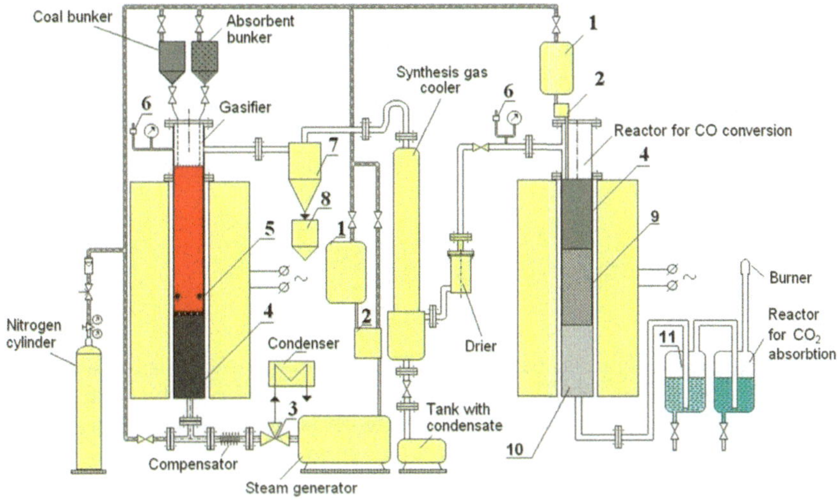

Figure 4.5 Experimental plant for solid fuel conversion into hydrogen and carbon acid gas.

burner. Feeders 2 fed water from tanks 1 into the steam generator and the reactor for CO conversion. Steam is fed from steam generator to the ceramic bed 4 in the gasifier through the three-way valve 3. The steam heated in the ceramic bed is fed into the reaction zone 5 of the gasifier. The obtained gas through the separator 7 is fed into the synthesis gas cooler. Unreacted solid residues are collected in the bunker 8. The cool gas, through the drier, is fed for heating into the ceramic bed of the reactor for CO conversion 4. In the reactor for CO conversion, there are beds of CO conversion catalysts: medium-temperature 9 and low-temperature 10. The obtained gas is fed into the reactor for CO_2 absorption through the cooler 11. Obtained hydrogen is combusted in the burner.

Standard methods of physical and chemical analysis of substances and own techniques developed in Thermal Energy Technology Institute (Coal Energy Technology Institute until 2021) were used in the work.

To expedite the component analysis of dry synthesis gas obtained in the process of char steam gasification, experimentally such formulae were deduced and checked to determine the volumetric concentrations of hydrogen and carbon dioxide in the dry gas:

$$[H_2] = 66.666 - 0.333 \cdot [CO] - 1.333 \cdot [CH_4], \tag{4.1}$$

$$[CO_2] = 33.333 - 0.666 \cdot [CO] + 0.333 \cdot [CH_4], \tag{4.2}$$

[CO] and [CH_4] are volumetric concentrations of CO and CH_4 in dry synthesis gas, %.

Equations (4.1) and (4.2) show that synthesis gas with volumetric concentrations of hydrogen to 66.7% and carbon dioxide to 33.3% ($C + 2H_2O = CO_2 + 2H_2$) can be obtained due to char steam gasification.

Reaction rate of fixed carbon was determined by the formula, kg/s:

$$R_c = \frac{12.011}{2240} \cdot V_{sg} \cdot ([CO]+[CO_2]+[CH_4]), \tag{4.3}$$

V_{sg} is the volume flow rate of dry synthesis gas, Nm^3/s.
The degree of carbon conversion was determined by the formula, %:

$$X_c = \frac{\sum_{i=1}^{n} R_{ci} \cdot \tau_i}{M \cdot \frac{C_c}{100}} \cdot 100, \tag{4.4}$$

R_{ci} is the reaction rate of carbon during the time τ_i, kg/s; M is the weight of char loaded in gasifier, kg; C_c is the mass content of carbon in the char, %.

It was deduced gross formula for char with steam interaction in gasifier taking excess steam during the process of reaction into account:

$$C + n \cdot \frac{[CO]+2\cdot[CO_2]}{s_c} H_2O = \frac{[CO]}{s_c} CO + \frac{[CO_2]}{s_c} CO_2 + \frac{[CH_4]}{s_c} CH_4 + \\ + \frac{[CO]+2\cdot[CO_2]-2\cdot[CH_4]}{s_c} H_2 + (n-1) \cdot \frac{[CO]+2\cdot[CO_2]}{s_c} H_2O \tag{4.5}$$

n is the ratio of steam and char consumptions ($n = 1$ for case of steam absence in obtained gas); $s_c = [CO] + [CO_2] + [CH_4]$, vol. %.

Volumetric concentration of components in wet synthesis gas was determined by the equations, vol. %:

$$[CO]_w = \frac{[CO]}{s_c \cdot m_{wc}} \cdot 100, \tag{4.6}$$

$$[H_2]_w = \frac{[H_2]}{s_c \cdot m_{wc}} \cdot 100, \tag{4.7}$$

$$[CO_2]_w = \frac{[CO_2]}{s_c \cdot m_{wc}} \cdot 100, \tag{4.8}$$

$$[CH_4]_w = \frac{[CH_4]}{s_c \cdot m_{wc}} \cdot 100, \tag{4.9}$$

$$[H_2O]_w = \left[(n-1) \cdot \frac{[CO]+2\cdot[CO_2]}{s_c \cdot m_{wc}}\right] \cdot 100, \tag{4.10}$$

$$m_{wc} = \frac{[CO]+[CO_2]-[CH_4]+n\cdot([CO]+2\cdot[CO_2])}{s_c}. \tag{4.11}$$

The degree of the use of steam was determined by the formula, %:

$$X_{H_2O} = \frac{[CO]+2\cdot[CO_2]}{n \cdot s_c} \cdot 100. \tag{4.12}$$

Lime bed was used over a char bed for CO_2 absorption in the char steam gasification reactor.

Reactions of lime, carbon, and steam in the gasifier were determined by such formulae:

$$CaO + H_2O \rightarrow Ca(OH)_2; \qquad \Delta H_{298^\circ} = -109\,kJ/mole, \tag{4.13}$$

$$C + H_2O \rightarrow CO + H_2; \qquad \Delta H_{298^\circ} = 132\,kJ/mole, \tag{4.14}$$

$$CO + H_2O \rightarrow CO_2 + H_2; \qquad \Delta H_{298^\circ} = -41{,}5\,kJ/mole, \tag{4.15}$$

$$Ca(OH)_2 + CO_2 \rightarrow CaCO_3 + H_2O; \qquad \Delta H_{298^\circ} = -69\ kJ/mole, \tag{4.16}$$

$$C + 2H_2O + CaO \rightarrow CaCO_3 + 2H_2; \qquad \Delta H_{298^\circ} = -88\,kJ/mole \tag{4.17}$$

ΔH_{298° is the thermal effect of reactions.

For estimation of hydrogen production processes, there were developed methods to determine: the degree of carbon dioxide absorption by lime in the char steam gasification reactor, the degree of CO conversion in the water-shift reactor, and the degree of carbon dioxide absorption in the low-temperature reactor (Dudnyk & Sokolovska, 2010).

Concentrations of obtained components before and after lime bed were determined at the initial stage of experimental research. The analysis of experimental data showed that, knowing the concentration of the obtained components in the dry synthesis gas after lime bed, the degree of carbon dioxide absorption by lime can be determined by calculation.

Concentrations of hydrogen and carbon dioxide in the obtained dry gas after char bed depend on concentrations of CO and CH_4 by formulae (4.1) and (4.2).

Concentrations of components in the synthesis gas before lime bed (on dry basis and without CO conversion in the lime bed) were calculated on the base of dependencies obtained by formulae (4.1) and (4.2) and synthesis gas composition after lime bed, %:

$$[H_2]^c = \frac{200}{3 + \frac{[CO]^{lm}}{[H_2]^{lm}} + 4 \cdot \frac{[CH_4]^{lm}}{[H_2]^{lm}}}, \tag{4.18}$$

$$[CO]^c = \frac{200 - 3 \cdot [H_2]^c}{1 + 4 \cdot \frac{[CH_4]^{lm}}{[CO]^{lm}}}, \tag{4.19}$$

$$[CH_4]^c = \frac{200 - 3 \cdot [H_2]^c}{\frac{[CO]^{lm}}{[CH_4]^{lm}} + 4}, \tag{4.20}$$

$$[CO_2]^c = \frac{100 - 2 \cdot [CO]^c + [CH_4]^c}{3}, \tag{4.21}$$

$[H_2]^{lm}$, $[CO]^{lm}$, $[CH_4]^{lm}$ are concentrations of hydrogen, CO, and methane in the dry synthesis gas after the lime bed, vol. %.

The degree of carbon dioxide absorption by lime, %:

$$X_{CO_2}^{L}=\frac{1-\dfrac{[CO_2]^{lm}}{[CO_2]^{c}}}{1-\dfrac{[CO_2]^{lm}}{100}}\cdot 100, \tag{4.22}$$

$[CO_2]^{lm}$ is the concentration of carbon dioxide in the dry synthesis gas after lime bed, vol. %.

Synthesis gas obtained after gasifier was fed into the water-shift reactor for CO conversion. CO conversion occurs in the presence of steam according to (4.15) on middle-temperature and low-temperature catalysts. The obtained mixture of gases consists mainly of hydrogen and carbon dioxide. The mixture was fed into the reactor for CO_2 low-temperature absorption to generate hydrogen.

As a result of a series of experimental studies, it was found that, knowing the concentration of gas components (on dry basis) before water-shift reactor and the presence of CO and CO_2 impurities in the obtained hydrogen (after the reactor for CO_2 low-temperature absorption), the degree of conversion of CO in water-shift reactor and the degree of CO_2 absorption in low-temperature reactor can be determined [6].

The degree of CO conversion in the water-shift reactor is determined by the formula, %:

$$X_{CO}=\left[1-\frac{[CO]^{out}}{[CO]^{in}}\cdot\left(1-\frac{[CO_2]^{in}-[CO_2]^{out}}{100-[CO_2]^{out}}\right)\right]\cdot 100, \tag{4.23}$$

$[CO]^{in}$, $[CO_2]^{in}$ are concentrations of CO and CO_2 in the dry synthesis gas before the water-shift reactor, vol. %; $[CO]^{out}$, $[CO_2]^{out}$, are concentrations of CO and CO_2 in the dry synthesis gas after the reactor for CO_2 absorption, vol. %.

The degree of CO_2 absorption in the low-temperature reactor, %:

$$X_{CO_2}=\frac{\dfrac{[CO_2]^{in}-[CO_2]^{out}}{1-\dfrac{[CO_2]^{out}}{100}}+X_{CO}\cdot\dfrac{[CO]^{in}}{100}}{[CO_2]^{in}+X_{CO}\cdot\dfrac{[CO]^{in}}{100}}\cdot 100, \tag{4.24}$$

The use of formulae (4.18)–(4.24) significantly reduced the time of analysis of processes of H_2 and CO_2 obtaining from renewable and fossil solid fuels.

Table 4.2 shows the results of analysis of solid fuels used in the studies.

The size of coal particles in the experiments was 0.63–1.0 mm.

Grind building magnesia lime was used in the experiments with coal. The size of lime particles was 0.4–1.6 mm. The content of active CaO in the lime was 85.1 wt.%, active MgO – 7.1 wt.%.

Table 4.2 Results of the fuel analysis.

Fuel	Proximate Analysis				Ultimate Analysis(daf)				
	Moisture	Volatile Matter	Ash Content	Fixed Carbon	C	H	O(dif)	N	S
	wt, %				wt, %				
Bituminous coal of Lviv–Volyn coal basin	6.45	28.95	26.8	37.8	81.6	5.4	10.2	1.5	1.3
Brown coal of Korostyshiv deposit of Zhytomyr region	11.06	36.47	25.60	26.87	63.6	5.8	26.0	0.8	3.8
Brown coal of Olexandria deposit of Kirovograd region	9.62	47.66	6.60	36.12	62.7	5.9	26.2	0.8	4.4
Sunflower husk char coal	5.9	41.4	32.3	32.3	75.8	3.8	20.1	0.3	0.0

Calcined dolomite was used in the experiments with sunflower husk charcoal. The size of calcined dolomite particles was 0.4–1.6 mm.

In the studies of volatile conversion, the experimental plant for solid fuel conversion into hydrogen and carbon acid gas consist of tank with distilled water, distilled water feeder, electric steam generator, reactor for steam conversion of volatile, cooler of synthesis gas, low-temperature reactor for CO_2 absorption with 25% water solution of monoethanolamine, burner, and measuring system for temperatures and gas concentrations.

In the mode of synthesis gas and hydrogen generation, the tank with distilled water, distilled water feeder, electric steam generator, steam gasifier, synthesis gas cooler, water-shift reactor, reactor for CO_2 absorption, burner, and measuring system for temperatures and gas concentrations were used in the experimental plant for solid fuel conversion into hydrogen and carbon acid gas.

Charcoal from sunflower husk was obtained in the process of carbonization in the plant for solid fuel conversion into hydrogen and carbon acid gas. 60 g of sunflower husks were loaded into the reactor. The rate of growth of the temperature of the layer up to 400°C was provided by electric heaters at a sample heating rate of 8–10°C/min. The obtained charcoal sample was carbonized for an additional 30 min at a temperature of 400°C.

4.3 RESULTS AND DISCUSSION

4.3.1 Steam conversion of Ukrainian bituminous coal (Lviv–Volyn coal basin) with hydrogen generation

4.3.1.1 Steam conversion of volatile matter

The reactor was loaded with 496 g of coal. The obtained products of bituminous coal volatile conversion were fed from the steam conversion reactor through the cooler into the reactor for CO_2 absorption. Figure 4.6 shows the results of bituminous coal volatile steam conversion and CO_2 absorption: ratio of components during the reaction (a), temperature in the bed of fuel (b), compositions of dry gases after the steam conversion reactor and absorption reactor (c), degree of CO_2 absorption (d). An increase in volumetric concentration of hydrogen in the dry gas up to 81.3% was reached due to the use of the reactor for CO_2 absorption.

4.3.1.2 Steam gasification of char using lime

The steam gasifier was loaded with char and lime (for capturing CO_2 in gasification zone directly). 221 g of lime was loaded at the top of the reactor. The weight ratio of carbon in the coke to CaO before the experiment was 4.72 g/g. Water feed rate in steam generator was 4.2 g/min.

Figure 4.7 shows the results of the steam char gasification: the degree of carbon conversion, gas feed rate before and after the bed of lime/limestone,

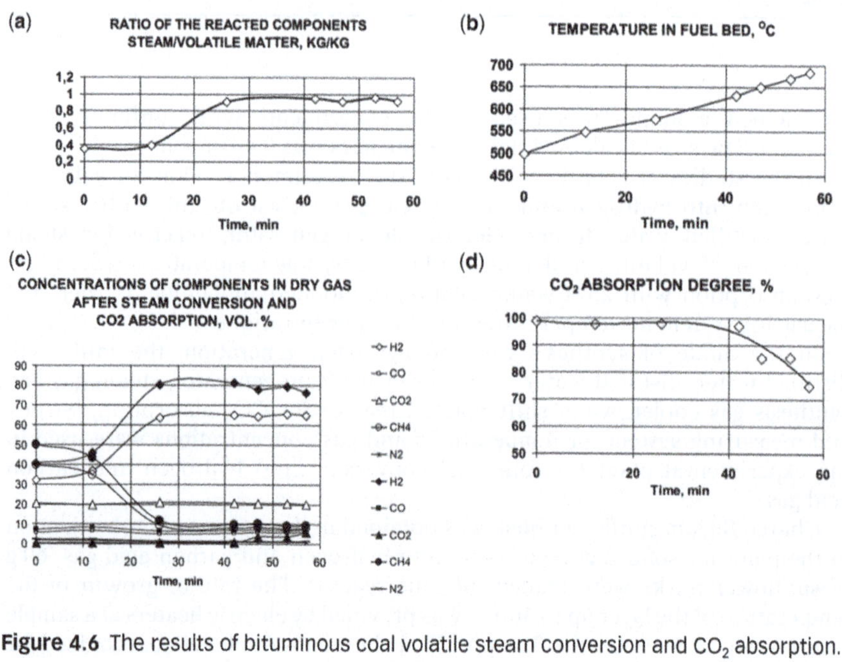

Figure 4.6 The results of bituminous coal volatile steam conversion and CO_2 absorption.

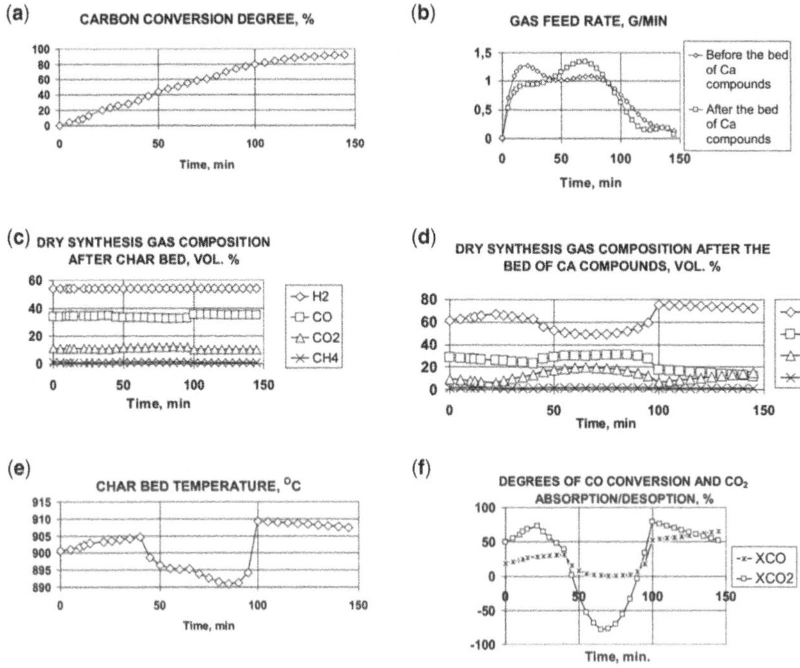

Figure 4.7 The results of steam gasification of Ukrainian bituminous coal char.

volumetric concentration of gaseous components before and after the bed of lime/ limestone, temperature in the bed of lime/limestone. Dry synthesis gas (H_2 = 72.4–75.2, CO = 11.5–17.3, CO_2 = 6.4–15.1, CH_4 = 0.97–1.00 vol.%) that is comparable by composition to synthesis gas after steam reforming of natural gas was obtained due to the use of lime in the steam gasifier of char.

At the final stage of the experiments, the synthesis gas was fed into the water-shift reactor. 171.8 g of ceramics (for generating steam from feed water and heating the steam-gas mixture), 401.5 g of iron-chromium catalyst (HTC-1), and 459.7 g of copper catalyst (LTC-4) were in the water-shift reactor.

Figure 4.8 shows the main operating parameters of the water-shift reactor: consumption of water in the reactor (a), the temperature in the middle of the water-shift reactor (b), the degree of CO conversion in the water-shift reactor (c), the composition of synthesis gas after the water-shift reactor (d). Synthesis gas was fed from the water-shift reactor through the cooler to the reactor for low-temperature CO_2 absorption. 110 ml of 25% monoethanolamine solution was in the reactor for low-temperature CO_2 absorption. The degree of CO_2 absorption in the low-temperature reactor was 95.6–97.8%.

Figure 4.9 shows the summary of results of char conversion into hydrogen at the last stage of the research.

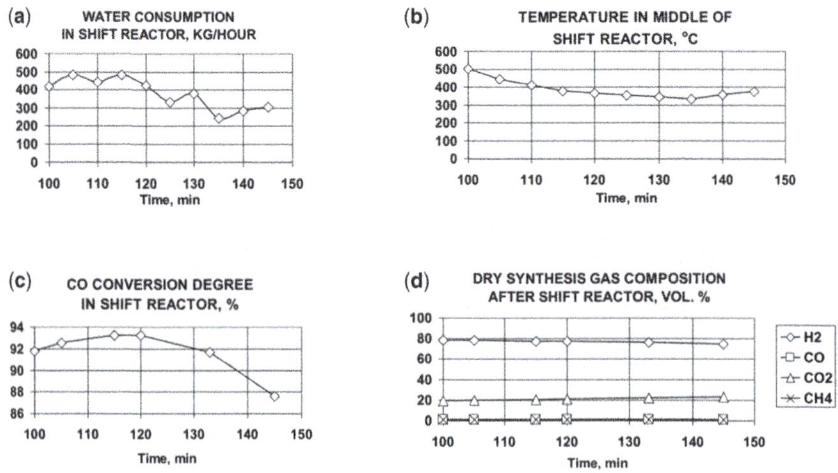

Figure 4.8 The main operating parameters of the water-shift reactor.

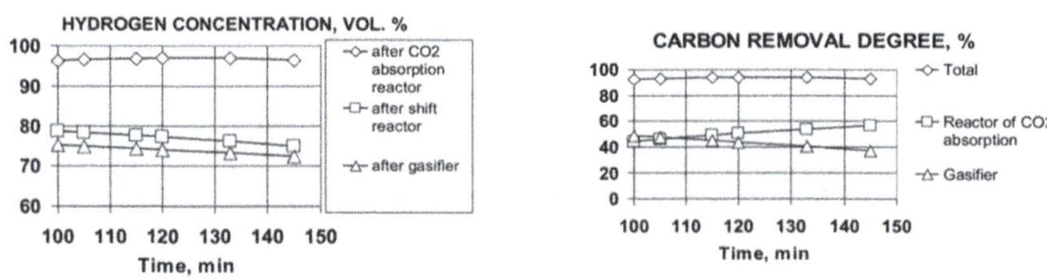

Figure 4.9 The summary of results of char conversion into hydrogen.

4.3.2 Steam conversion of Ukrainian brown coal (Korostyshiv deposit of Zhytomyr region)

4.3.2.1 Steam conversion of volatile matter

A measure of 50 g of brown coal was loaded into the reactor for steam conversion of volatile to produce char. Figure 4.10a shows the steam consumption for the steam conversion of brown coal volatile in the process of the experiments. Figure 4.10b shows the temperature in the coal bed and in the middle of space over the bed. Figure 4.10c shows the composition of the obtained dry gas.

During the volatile steam conversion, the hydrogen content in the obtained gas changed from 11.5 to 52.7 vol. %, methane content – from 47.9 to 19.7 vol. %, CO – from 7.4 to 15.2 vol. %, CO_2 – from 2.4 to 12.4 vol. %.

The result of proximate analysis of the obtained char is: volatile – 6.5 wt. %, ash – 45.38 wt. %, moisture – 1.67 wt. %.

4.3.2.2 Steam gasification of char using lime and catalyst of CO conversion

A measure of 46 g of lime, 50 g of iron-chromium catalyst CTK-1, and 46 g of lime were loaded in consecutive order into the gasifier on the bed of the obtained char.

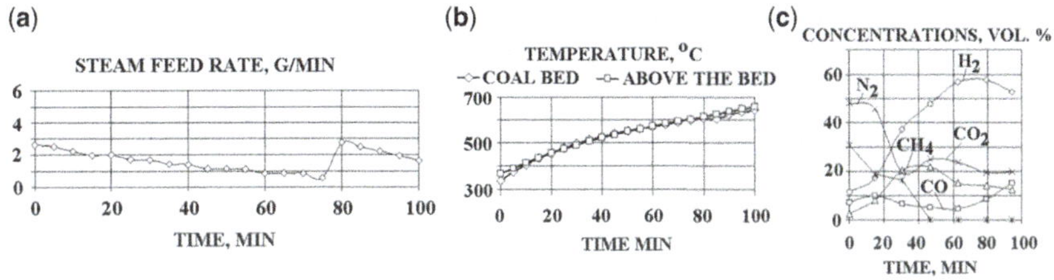

Figure 4.10 Steam conversion of volatile matter.

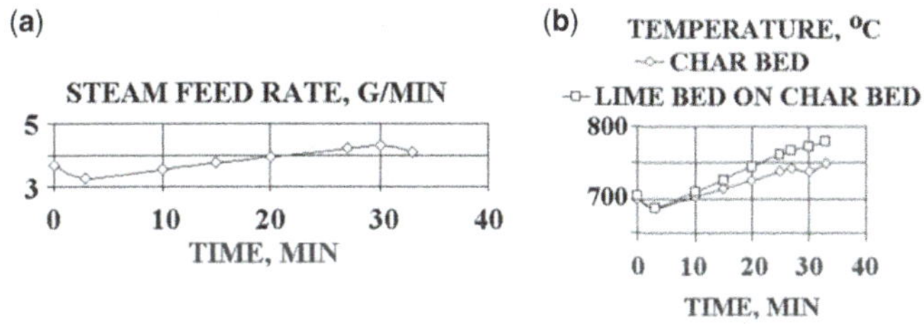

Figure 4.11 Steam consumption in char gasification process (a) and temperature in the char bed and temperature of lime on the char bed (b).

Figure 4.11a shows the steam consumption in char gasification process, Figure 4.11b shows the temperature in the bed of char and temperature of lime/limestone on the char bed.

Figure 4.12 shows the consumption of dry synthesis gas and its composition at the gasifier outlet (a, b), the reaction rate of carbon (c), and degree of carbon removal in the gasifier (d).

Figure 4.12b shows that technical hydrogen (gas containing over 80 vol. % of hydrogen) was obtained at the plant during the first 22 min of experiment, after which the hydrogen concentration decreased to 67.4 vol. %. During the first 20 min of the experiment, the CO concentration in the obtained gas was 1.66–3.25 vol. %, after which it increased to 5.69 vol. %.

During 34 min of experiment, the absorbent removed 59.27% of carbon, which has reacted with steam.

4.3.3 Steam conversion of Ukrainian brown coal (Olexandria deposit of Kirovograd region) with hydrogen generation

4.3.3.1 Steam conversion of volatile matter

A weight of 50 g of brown coal was loaded into the reactor for steam conversion of volatile to produce char.

Figure 4.13a shows the steam consumption for the steam conversion of brown coal volatile. Figure 4.13b shows the temperature in the coal bed and in the middle of space over the bed. Figure 4.13c shows the composition of the obtained dry gas.

During the volatile steam conversion, the hydrogen content in the obtained gas changed from 6.8 to 66.6 vol. %, CO_2 – from 14.5 to 55.1 vol. %, methane content – from 3.1 to 16.6 vol. %, CO – from 2.7 to 13.8 vol. %

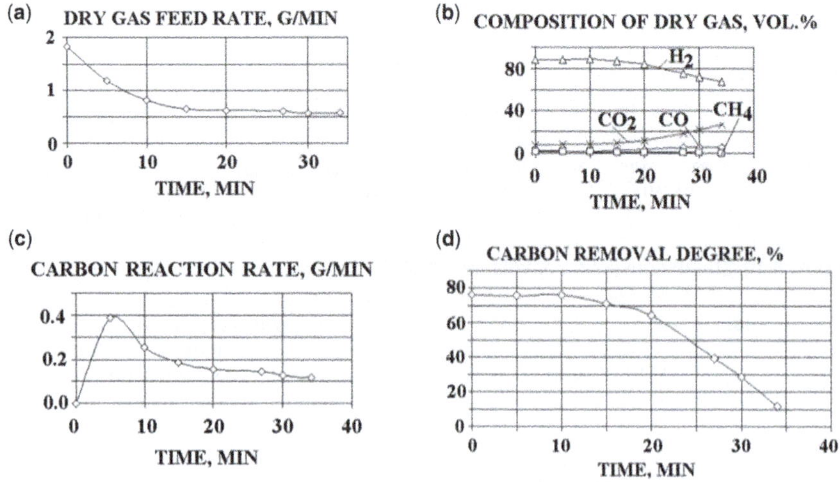

Figure 4.12 Results of steam conversion of char using lime and CO catalyst in the gasifier: consumption of dry synthesis gas and its composition (a, b), the reaction rate of carbon (c), and degree of carbon removal in the gasifier (d).

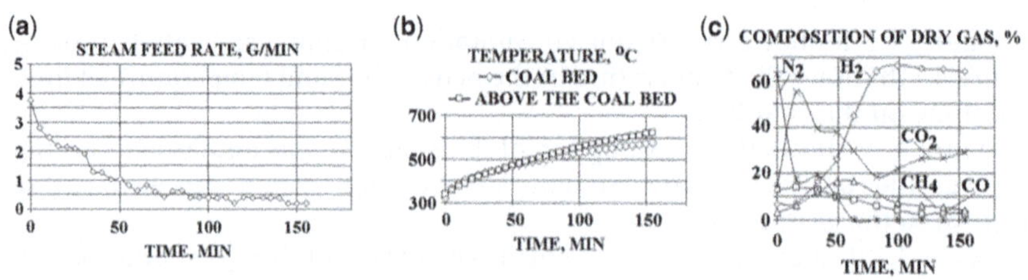

Figure 4.13 Steam conversion of volatile matter.

The result of proximate analysis of the obtained char is: Volatile – 13.84 wt. %, Ash – 14.13 wt. %, Moisture – 9.1 wt. %.

4.3.3.2 Steam gasification of char using lime and hydrogen generation

124.6 g of CO_2 absorbent (lime) were loaded into the gasifier on the bed of the obtained char. Figure 4.14a shows the steam consumption in char gasification process, Figure 4.14b shows the temperature in the bed of char and absorbent.

Figure 4.15 shows the consumption of dry synthesis gas and its composition at the gasifier outlet (a, b), the reaction rate of carbon (c) and degree of CO_2 absorption in the gasifier (d).

Figure 4.15b shows that hydrogen concentration in the dry gas was 82.4–84.1 vol. % during the first 15 min. of experiment, after which it decreased to 66.1 vol. %. The concentration of CO changed in a range from 7.1 to 3.7 vol. %, CO_2 – from 7.7 to 29.7 vol. %, CH_4 – from 1.9 to 0.5 vol. %.

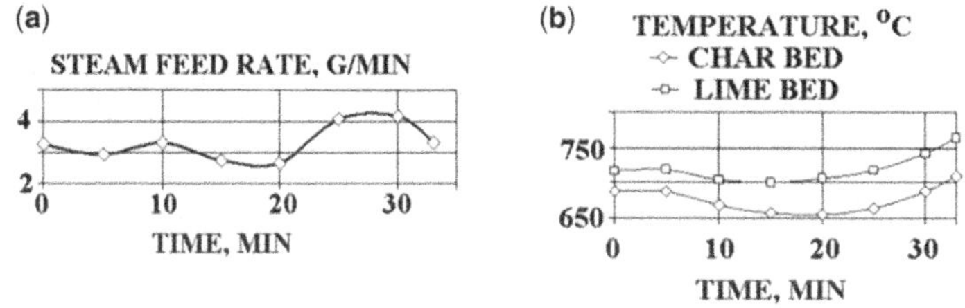

Figure 4.14 Steam consumption in char gasification process (a) and temperature in the bed of char and absorbent (b).

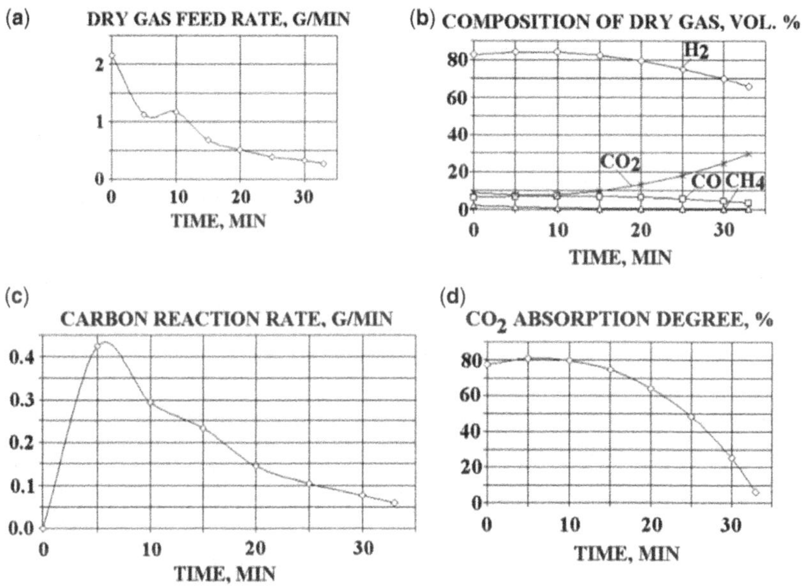

Figure 4.15 Results of steam conversion of char using lime in the gasifier: consumption of dry synthesis gas and its composition (a, b), the reaction rate of carbon (c) and degree of CO_2 absorption (d).

Figure 4.15d shows that degree of carbon dioxide removal in the gasifier was 74.5–80.8% during the first 15 min of gasification. Then, the degree of CO_2 absorption decreased to 6.3%. During 33 min of experiment, the absorbent removed 51.8% of carbon, which has reacted with steam.

Dry synthesis gas obtained after gasifier was fed through the shift reactor to the reactor for low-temperature absorption of CO_2. The loading of the shift reactor was the same as in the case of conversion of bituminous coal char into hydrogen. Figure 4.16 shows the degree of CO conversion in the water-shift reactor (a); volumetric concentration of hydrogen at the gasifier outlet, water-shift reactor inlet, and outlet of the reactor for CO_2 absorption (b); hydrogen

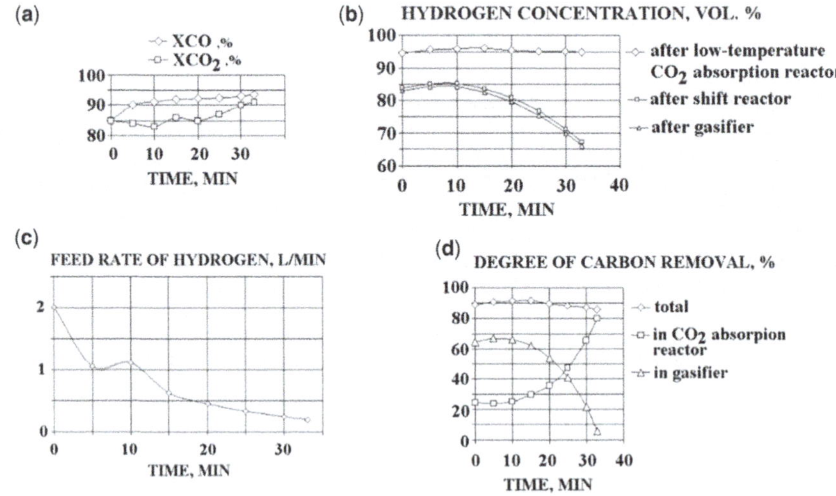

Figure 4.16 Results of char conversion into hydrogen: the degree of CO conversion in the water-shift reactor and the degree of CO_2 removal in the reactor for low-temperature absorption of CO_2 (a); volumetric concentration of hydrogen in dry gas at the gasifier outlet, water-shift reactor inlet, and outlet of the reactor for CO_2 absorption (b); hydrogen consumption at the outlet of the reactor for low-temperature absorption of CO_2 (c); the degree of carbon removal in the gasifier, reactor for low-temperature absorption of CO_2, and the total degree of carbon removal in the installation (d).

consumption at the outlet of the reactor for low-temperature absorption of CO_2 (c); and the degree of carbon removal in the gasifier, reactor for low-temperature absorption of CO_2, and the total degree of carbon removal in the installation (d).

The degree of carbon removal was: in the gasifier – 5.6–67.1%; in the reactor for low-temperature absorption of CO_2 – 23.7–80.2%. The degree of carbon removal in the installation was 85.8–91.7%. As a result, there was obtained hydrogen purity of 94.7–96.2%.

89.7% of carbon, which reacted with steam in the gasifier, was removed during all time of experimental researches in the gasifier and reactor for low-temperature absorption of CO_2.

4.3.4 Steam gasification of sunflower husks charcoal using Ni-catalyst and calcined dolomite

The yield of dry coal from dry weight of sunflower husk was 34.5%. 20.7 g of charcoal was loaded into the reactor in the gasifier. The composition of charcoal was, wt. %: $A^d = 21.7$, $V^d = 44.0$, $C_f^d = 34.3$. The elemental composition of dry charcoal, wt.%: $C^d = 59.5$, $O^d = 15.8$, $H^d = 3.0$, $A^d = 21.7$. LHV of charcoal was 21.4 MJ/kg.

The charcoal bed was loaded with: Ni catalyst – 8 g, calcined dolomite – 65.0 g. Catalyst composition was, wt. %: NiO – 5.5, Al_2O_3 – 84.5%. The size of the calcined dolomite particles was 0.4–1.6 mm. The catalyst was dried at the temperature of 110°C for 1 hour and regenerated in a flow of a mixture of hydrogen and argon (15/85 wt. %) at a temperature of 450°C for 2 h.

Figure 4.17 shows the steam feed rate into gasifier (a), temperature of fuel bed (b), dry gas yield (c), and composition of obtained dry gas (d) during catalytic steam gasification with the use of Ni-catalyst and CO_2 capture by calcined dolomite. The temperature in the coal bed was maintained at 700–770°C. The temperature inside the bed of solid products of lime conversion was 525–580°C.

During the steam gasification of charcoal from sunflower husks with the CO_2 capture, the concentrations of components in dry gas were, vol. %: H_2 – 62,1…72.9 vol. %; CO_2 – 9.2…13.4 vol. %, CH_4 – 1.9…2.7, CO – 11.0…20.9 vol. %. The degree of CO_2 capture by calcined dolomite was 60.1…68.9%.

Figure 4.18 shows the degree of carbon conversion of charcoal during catalytic steam gasification with CO_2 capture by calcined dolomite. After steam gasification, 98.9% of carbon was converted.

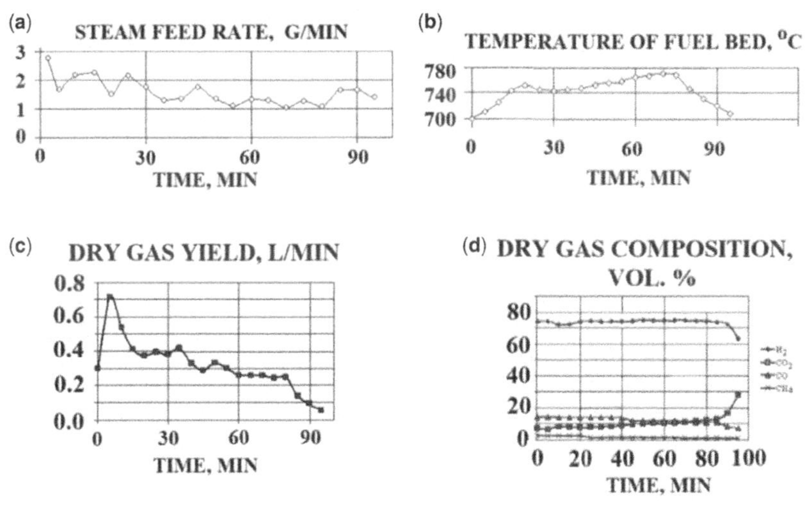

Figure 4.17 Steam consumption in the steam generator (a), temperatures inside the bed of charcoal (b), yield of dry gas after gasification and CO_2 capture (c), dry gas composition after steam gasification and CO_2 capture (d).

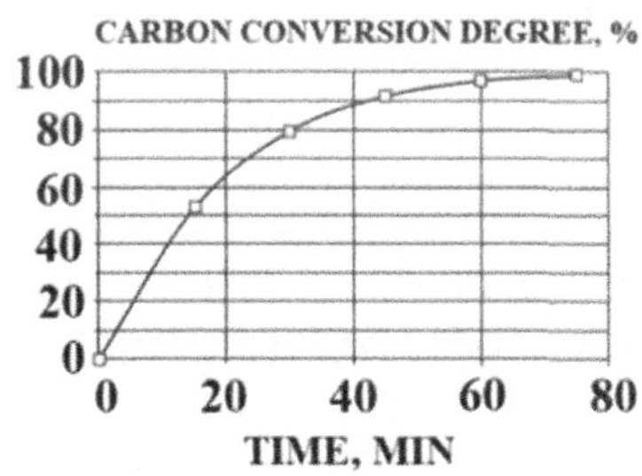

Figure 4.18 Degree of carbon conversion of charcoal during catalytic steam gasification with CO_2 capture by calcined dolomite.

4.4 CONCLUSIONS

(1) Today, there is a need to replace natural gas and liquid fuels with alternative fuels for the production of energy and chemical products. In Poland, most electricity is generated using coal.

(2) The composition of fuels and efficiency of fuel use are determining factors that have an impact on the carbon dioxide in recalculation on kWh of generated energy. An increase in hydrogen content in the fuel organic mass leads to a decrease in carbon dioxide yield in recalculation on kWh of generated energy.

(3) In the world, most technologies for the production, purification, and utilization of carbon dioxide have already been tested in various industries on a commercial scale. The lowest cost of CO_2 removal in solid fuel conversion processes is achieved using IGCC power plants. Reduced cost and improved efficiency of CO_2 sequestration are provided by removing CO_2 not from combustion products, but from solid fuel gasification products – synthesis gas.

(4) Thermal Energy Technology Institute of NAS of Ukraine carries out research on the production of synthesis gas, H_2, and CO_2 from coal, biomass, and solid organic waste. The possibility of using CO_2 for the production of value-added products using renewable energy sources is being analyzed.

(5) An experimental plant for the production of H_2 and CO_2 from coal, biomass, and solid organic waste has been created.

(6) Up to 50% and more of carbon is captured during catalytic steam gasification of bituminous and brown coals (using lime), biomass and solid organic waste (using calcined dolomite). After the first stage of CO_2 capture, hydrogen-rich synthesis gas is fed to the catalytic water shift reactor for CO conversion. After the shift reactor, the resulting mixture of hydrogen and CO_2 is fed to the low-temperature CO_2 sorption reactor with an aqueous solution of monoethanolamine. The main products of the experimental plant are hydrogen and carbon dioxide.

(7) Dry synthesis gas ($H_2 = 72.4–75.2$, $CO = 11.5–17.3$, $CO_2 = 6.4–15.1$, $CH_4 = 0.97–1.00$ vol.%), which is comparable by composition to synthesis gas from steam reforming of natural gas, was obtained from char of bituminous coal due to the use of lime. Hydrogen of 97% purity was obtained with the use of steam gasifier, water-shift reactor, and low-temperature CO_2 absorption reactor. 92–94% of carbon was removed in the system of hydrogen generation: 37–48% – in the gasifier, 44–56% – in the low-temperature CO_2 absorption reactor.

(8) Char of brown coal was used to obtain hydrogen by two circuits: (1) steam gasifier with middle-temperature catalyst of CO conversion inside the bed of lime; (2) steam gasifier (with lime), shift reactor (with catalysts of CO conversion) and CO_2 absorption reactor (with monoethanolamine solution).

(9) Hydrogen of 95–96% purity was obtained from char of brown coal as a result of the use of steam gasifier with lime, shift reactor with

iron-chromium and copper catalysts of CO conversion, and low-temperature CO_2 absorption reactor with monoethanolamine solution.

(10) In the case of the use of brown coal char at the temperature of steam gasification less than temperature steam gasification of bituminous coal char by 230 °C, maximal volumetric concentration of hydrogen in the obtained dry gas after the gasifier was increased by 9% (up to 84.3%) with increasing carbon removal in the gasifier.

(11) As a result of steam gasification of coal (obtained from sunflower husks) at temperatures of 745–770°C using beds of Ni-catalyst and calcined dolomite, hydrogen-rich dry synthesis gas containing, wt. %: $H_2 = 72.5$; $CO = 13.6$; $CH_4 = 2.5$; $CO_2 = 11.4$. The degree of absorption of CO_2 by absorber was 64.5%, the LHV of the obtained synthesis gas was 10.4 MJ/Nm^3.

(12) In case of the use of lime or calcined dolomite for CO_2 capture, consumption of the water in low-temperature system of CO_2 capture was decreased.

(13) The main products of the experimental plant are hydrogen and carbon dioxide. The innovative nanoparticle ferriferrohydrosol (FFH) to capture CO_2 and produce hydrogen can be used at the experimental plant.

ACKNOWLEDGMENTS

Scientific work in framework of Target complex Programme for Research of NAS of Ukraine "Development of scientific bases for hydrogen production, storage and use in autonomous energy supply systems" was carried out.

REFERENCES

Beginning the World's First Integrating Coal Gasification Fuel Cell Combined Cycle Project (IGFC) Demonstration Project–Aiming to both Improve Efficiency of Coal-Fired Power Generation and Greatly Cut CO_2 emissions. News Release, NEDO, OSAKI CoolGen Corporation. April 17, 2019, p. 4.

BP Statistical Review of World Energy (2022). 71st edition, p. 60.

CO_2 Power Solutions. Extra power with no additional fuel burned. E Marine. A division of GE Aviation, p. 2.

Dudnik A. N. (2012). Concepts for new thermal coal-fired power plants with carbon dioxide removal and hydrogen production. *Modern Science: Research, Ideas, Results, Technologies. Collection of Scientific Articles*, **3**(11), 91–97.

Dudnyk O. M. (2014). Development of processes and flow sheets for the conversion of Ukrainian coal, woody biomass and organic waste of energy sector and chemical industry. Advanced technologies developed by Coal Energy Technology Institute of National Academy of Sciences of Ukraine for Thermal Power Industry, Chapter: 5.3. "GNOZIS" Ltd., Kyiv, Ukraine, pp. 189–214.

Dudnyk O. M. and Sokolovska I. S. (2010). Conversion of Ukrainian low grade solid fuels with CO_2 capture. Proc. of 27th Ann. Int. Pittsburgh Coal Conference 2010 (PCC 2010), Istanbul, Turkey, Oct. 11–14, vol. **2**, pp. 1012–1033.

Dudnyk O. M., Dunaevskaya N. I., Sokolovska I. S., Tripol'skii A. I. and Strizhak P. E. (2020). CO_2 production, purification and utilization in the processes of power generation and chemical production. Proc. of XVI Int. Science and Practice Conf. on Thermal Energy: Ways of Reconstruction and Development, November 13, Kyiv, pp. 46–55.

Dudnyk O. M., Ostapchuk M. V., Sokolovska I. S. and Zhovtyansky V. A. (2025). Plasma steam-air conversion of solid biomass and problem of equilibrium assumption in the gasification process. *Energotehnologii i Resursosberezenie*, **84**(3), 3–20.

Esposito R. (2017). The Kemper Project IGCC. Project Overview. SECARB Stakeholders Briefing.

Herbst M. (2015). Biogas-fired Combined Hybrid Heat and Power Plant for sustainable energy. European Union's Horizon 2020 Research and Innovation Program. No 641073. 2 p.

Official Poland – Country Commercial Guide (2022-07-22). Website of the International Trade Administration. https://www.trade.gov/country-commercial-guides/poland-energy-sector (accessed 14 July 2022).

Wakabayashi Y. (2017). The Latest Coal-Fired Thermal Power Plant. Mitsubishi Hitachi Power Systems, p. 38.

Wang T. and Stiegel G. J. (2016). Integrated Gasification Combined Cycle (IGCC) Technologies, 1st edn. Woodhead Publishing, Cambridge, UK, p. 928.

Xiaodong P. (2018). Cleaning, upgrading, storing technology of super biological naturalgas plants and typical cases. Proc. of Seminar on Bio-natural Gas for Developing Countries, September 7–27, Chengdu.

Zhovtyansky V., Dudnyk O., Petrov S., Verbovsky V., Rubets D. and Yakymovych M. (2013). Plasma-steam intensification of the hydrogen yield in the process of wood air gasification. *High Temperature Material Processes–An International Quarterly of High-Technology Plasma Processes*, **17**(1), 25–35.

Zhovtyansky V. A., Dudnyk O. M., Ostapchuk M. V. and Sokolovska I. S. (2023). Conversion of carbon-containing raw material using plasma technologies. *Energotehnologii i Resursosberezenie*, **77**(4), 90–112.

Zhu Q. (2019). Developments on CO_2-utilization technologies. *Clean Energy*, **3**(2), 85–100.

doi: 10.2166/9781789065138_0055

Chapter 5

Landfill gas calorific value enhancing by carbon dioxide extraction to boost power generator efficiency

H. Zhuk[1], Yu Ivanov[1], S. Krushnevych[1], V. Verbovskiy[1], D. Komissarenko[2] and M. Bondarenko[2]*

[1]The Gas Institute of the National Academy of Science of Ukraine, Dehtiarivska Street 39, Kyiv, Ukraine 03113
[2]International Centre of Gas Technologies LLC, Ivana Honty Street 1, Kyiv, Ukraine 04112
**Corresponding author: hen_zhuk@ukr.net*

ABSTRACT

Ukraine hosts over 6000 landfills for municipal solid waste, annually receiving up to 12 million tons of waste. During the anaerobic digestion process, these landfills emit approximately 700,000 to 800,000 tons of methane and 1,800,000 tons of carbon dioxide each year. While landfills serve as a significant source of biogas, they also pose a genuine environmental challenge.

Ukrainian landfills in cities such as Kyiv, Odesa, Sumy, Kharkiv, Ivano-Frankivsk, Lviv, Mykolaiv, and Chernihiv were thoroughly investigated to study biogas emissions. The primary method of biogas utilization in Ukraine involves electricity production (with a total installed capacity of 100 MW) and its supply to the power grid.

The International Center for Gas Technologies LLC, collaborated with the Gas Institute of the National Academy of Sciences of Ukraine, successfully implemented a project to collect landfill biogas and produce 0.3 MW electric power at the Kamyanets-Podilsky landfill, which contains about two million tons of solid waste.

Through computer simulations, researchers developed technological schemes for amine-based carbon dioxide extraction from biogas. An effective MDEA-modified absorbent was proposed, significantly reducing the load on the desorber reboiler compared to traditional MEA solutions.

During the intensive extraction of landfill gas, methane concentrations can drop to 30% or lower. From other side, gas piston engines influenced by a critical methane content of 28%. To solve this challenge, researchers investigated biogas enrichment (via carbon dioxide extraction) while maintaining an increased air content (20–40% of nitrogen). This approach led to 1.2–1.4 times increasing in methane concentration, with the heat consumption amounting to 1.0 kW per 1 kg of carbon dioxide extraction across the entire range of biogas compositions.

Keywords: landfill gas, piston engine, power generation, amine absorption, carbon dioxide extraction, computer simulation

5.1 INTRODUCTION

The Gas Institute of the National Academy of Science of Ukraine was founded in 1949 in Kyiv, Ukraine. The main research directions are natural and oil gas processing, transportation and use, highly effective gas burning, engine conversion to natural gas, synthesis gas, biogas, alternative fuels use, natural gas liquefaction, and so on. In 2009 the Gas Institute began research in the field of collecting and using biogas on landfills by the initiative of Deputy Director and team leader of the gas processing department, Alexandr Pyantichko (1936–2020).

There are more than 6000 landfills of municipal solid waste in Ukraine, and up to 12 million tons of waste are filled annually. In the process of anaerobic digestion, they emit about 700–800 thousand tons of methane and 1800 thousand tons of carbon dioxide per year. Landfills are a real environmental problem but they can be a significant source of biogas too. The simplest way to reduce damage to the environment from greenhouse gases is burning methane because it has near 81 times in short-term (GWP20) and 28 times in long-term (GWP100) higher greenhouse effect compared with carbon dioxide (AR6 WGI Report, 2021). Today biogas is widely used at power stations in Ukraine. Nearly 90 landfills have enough capacity and structure of waste for commercial efficiency electricity production in Ukraine and 26 already have electricity production with 100 MW total installed power and producing 112.3 GWh per year. It means the effectiveness of installed power is below 50%, mainly because of the low calorific value of extracted landfill gas.

Fresh waste begins to generate methane a few years after being imported into a landfill and can be a stable source of methane for up to 40 years (Pyatnichko *et al.*, 2013; Tashyrev *et al.*, 2022). After that period anaerobic process speed decreased significantly and methane concentration was not enough to use in engines as primary fuel. When methane concentration decreased the full load power and stability of the engine also decreased, because of the increased amount of inert gases and oxygen which the engine must push through itself and the calorific value of fuel decreased. For example, the biogas engine Caterpillar G3508 has 28% (determined experimentally) as the minimum allowed concentration of biomethane for stable work.

Biogas not only works as a heat source; it is also a commercial carbon dioxide source. By biogas separation it is possible to increase engine efficiency and allow the generation of commercial products which have a large number of consumers including fire stations (including extinguishing fires in landfills) and greenhouses. So, carbon dioxide separation is the next stage in biogas utilization.

Figure 5.1 Research of flow rate and gas content of the well.

5.2 LANDFILL GAS RESEARCH

The first stage of commercial biogas utilization is landfill research. The Gas Institute specialists made research wells on 17 landfills in Ukraine of different age wastes with special equipment emulated really well debit, Figure 5.1, see Pyatnichko *et al.* (2013) and Zhuk *et al.* (2024). Gas probes were taken for analysis of gas content changing over time.

We analyzed obtained gas samples in our laboratory using the Agilent 6890N gas chromatograph. Methane content in biogas exceeded 50% so it can be easily used directly in engines. Also, biogas includes 20–40% of carbon dioxide, and less than 6.5% of nitrogen (Figure 5.2).

For the process of biomethane extraction from biogas simulation software CasCondOil and Aspen HYSYS were used, and optimized flowcharts and technical parameters for certain conditions are determined.

The first implementation of biogas harvesting and use at a power station with a 1 MW installed power engine in Ukraine was made on a municipal landfill near Kyiv in 2012, Pyatnichko *et al.* (2013). In 2015 the installed power was increased here to 2 MW. Up to today, we participated in the implementation

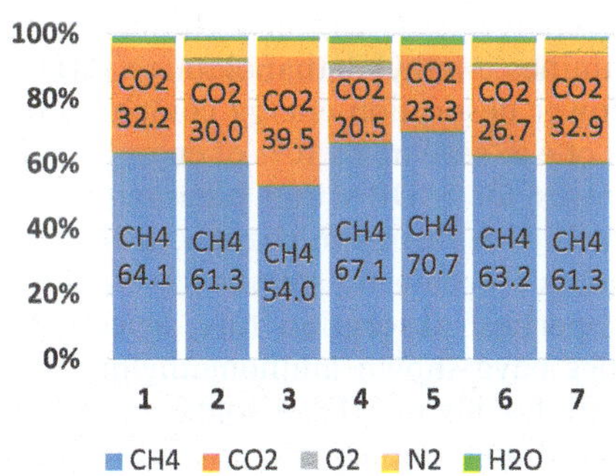

Figure 5.2 Results of different landfills biogas analysis (average data).

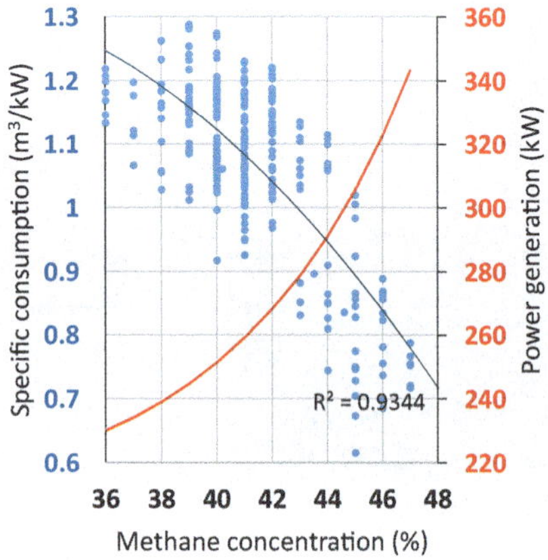

Figure 5.3 Effect of methane concentration on specific gas consumption (blue points) and power generation with 300 m^3/h constant biogas consumption (red curve).

of biogas utilization on six landfills in Ukraine with a 6.5 MW total installed power. On one of the implemented landfills at Kamyanets-Podilskyi town (operating company the International Center of Gas Technologies LLC) with a 500 kW power engine Caterpillar G3508 we made detailed data measurements through the last 2 years. It allows us to build charts of different parameters changing including specific consumption of fuel with methane concentration change. Also, electricity power change with constant biogas flow (300 m^3/h) with methane concentration change was built (Figure 5.3).

5.3 RESULTS AND DISCUSSION

Operating costs of carbon dioxide extraction from biogas significantly consist of energy consumption for the regeneration of a saturated solution, it is up to 70% for the amine process.

The development of absorbents that provide an optimal compromise between fast kinetics and low energy costs can be obtained by introducing activated additives or mixing amines. Methyldiethanolamine (MDEA) with the addition of piperazine has a lower desorption energy compared to monoethanolamine (MEA), which allows one to save up to 10% of energy for solution regeneration, while evaporation losses for MDEA and piperazine are significantly lower than for MEA (Anuchin & Miroshnichenko, 2015; Saleem *et al.*, 2017). Aminomethylpropanol is a promising solvent for carbon dioxide removal due to its absorption properties: absorption rate, selectivity, and resistance to decomposition. Studies have shown aminomethylpropanol effectively affects the solubility of carbon dioxide in MDEA solution, and the rate of absorption increases with increasing aminomethylpropanol concentration in water MDEA solution (Al-Masabi & Castier, 2011; Gang *et al.*, 2017). Capraa *et al.* (2018) suggested using a mixture of MDEA/MEA with a concentration of each component of 20% (wt.) as an absorbent (Table 5.1).

Table 5.1 Indicators of biogas purification process by different amine sorbents.

	Composition of the Absorbent, % (wt.)					
	15%MEA	18%MEA	20%MEA	7%MIEA + 30%MDEA	8%MEA + 40%MDEA	MDEAmod
Absorbent flow rate, kg/h	4795	4220	3940	4380	3730	3100
Reboiler load, kW	375	405	425	200	180	165

Calculations of energy costs in the production of biomethane as engine fuel using amine and water processes for the extraction of carbon dioxide from biogas are mentioned in Ivanov *et al.* (2017), (2021).

The combined water–amine method for the scrubbing of biogas from carbon dioxide was proposed in Capraa *et al.* (2018). The preliminary extraction of carbon dioxide from biogas occurs with the water absorption at a pressure close to atmospheric or increased to 0.2–0.3 MPa. For the final scrubbing of biogas to biomethane, the amine absorption was used.

5.3.1 Modelling of the water absorption process

The process of water scrubbing of biogas from carbon dioxide takes place with the use of freshwater that does not contain impurities and circulates in the circuit. Water from the city water networks, wells, or prepared rainwater can be used as fresh water.

Figure 5.4 shows a two-stage technological flowchart for water scrubbing of biogas from carbon dioxide using two options.

5.3.1.1 Brief description of the technological process of preliminary scrubbing of biogas from carbon dioxide

Biogas supplied to the scrubbing column (absorber A-1) at a pressure 0.26 MPa, because higher pressure leads to an increase in the energy consumption of the process and to significant losses of methane due to its dissolution in water. The freshwater used as absorbent is supplied to the absorber counterflow to the gas. In this process, carbon dioxide dissolves in water, and enriched methane leaves the absorber for further amine scrubbing at a pressure 0.26 MPa.

Carbon dioxide saturated water leaves the absorber and supplies it to the evaporation column K-1, where it is blown by air under a pressure of 0.12–0.13 MPa, while the dissolved methane is almost completely removed from water. The degassing gas (aeration gas) which leave the evaporation column K-1 can be mixed with the biogases fed flow to the scrubbing column A-1 (variant 'A') to increase the methane content at the outlet of the column, or used as fuel gas (variant 'B') with methane concentration more than 30%. In the case of a methane concentration less than 30%, it is impractical to use this gas as a fuel. In the evaporative column K-2, at close to atmospheric pressure, the water solution is blown by counterflow air, while carbon dioxide is almost completely removed from the absorbent (water), which is then supplied again to the absorber A-1.

The GasCondOil software was used to simulate water absorption processes.

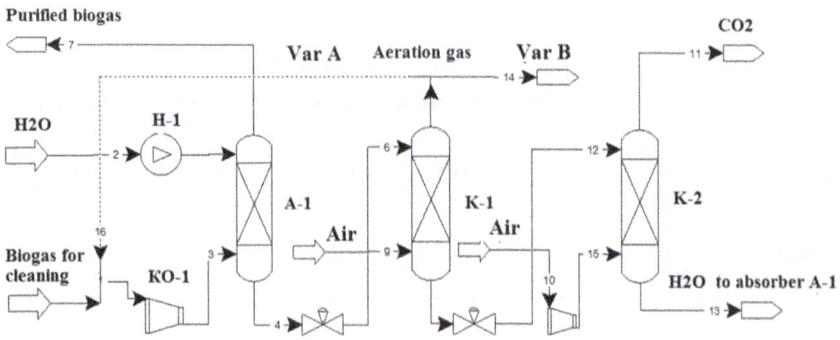

Figure 5.4 Two-stage technological flowchart of water scrubbing of biogas from carbon dioxide. A-1 – absorber; K-1, K-2 – evaporation columns; KO-1 – biogas compressor; H-1 – water pump.

5.3.2 Modelling of the amine absorption process

Figure 5.5 shows the basic technological flowchart of biogas scrubbing by water solutions of amines, with biomethane and carbon dioxide obtained. Software Aspen HYSYS was used to simulate these processes.

Biogas at P = 0.26 MPa and T = 313 K (40°C) supplied to absorber A-1. The absorber is irrigated with a water solution of the chemisorbent with temperature T = 318 K (45°C). In the absorber column, the concentration of carbon dioxide in the biogas decreases to 2% (vol.). Enriched biogas (biomethane) is sent to the consumer. The saturated chemisorbent solution enters the recuperative heat exchanger TO-1, in which it is heated to a temperature of 373 K (100°C) by the hot reverse flow of the regenerated chemisorbent solution, which comes from the desorber (stripper column) D-1. The heated saturated chemisorbent solution is supplied to the upper side of the desorber D-1, where the absorbed carbon dioxide is vaporized to the required concentration. The regeneration process is carried out using a reboiler column Reb at a chemisorbent boiling point of 380–391 K (114–118°C). The vapor–gas mixture that comes out of the upper side of the desorber is cooled in the dephlegmator Def to 298–303 K (25–30°C), while the water vapor condenses and is supplied to the desorber as irrigation in its upper side, and the gas that leaves the dephlegmator mainly contains 98% vol. of carbon dioxide. The regenerated chemisorbent solution is supplied by pump H-1 into the recuperative heat exchanger TO-1, then into the cooler OH-1, and thereafter the cooler is supplied into the upper side of the absorber A-1.

5.3.3 Results of comparative calculations of water, amine, and combined water–amine absorption processes

The results of the simulation of water, amine, and combined water–amine carbon dioxide absorption processes of 250 nm³/h of biogas of different compositions are presented in Table 5.2. The carbon dioxide concentration in biogas varied from

32 to 42% (vol.). Such concentrations of carbon dioxide are the most common for biogas, in particular, biogas from household waste landfills (30–40%).

According to the variant 'A' flowchart (Figure 5.4), where the degassing gas of the evaporating column K-1 is mixed with biogas in order to increase the potential content of methane at an output and is fed to the input of the compressor KO-1, the combined water–amine absorption has an excess of energy costs by 50% compared to water absorption, and methane losses are insignificant (0.14–0.24%).

When using variant 'B' (Figure 5.4) in the aeration gas of column K-1, the concentration of methane is, on average, about 35%, and this gas can be used as a fuel for the production of both electricity and heat and, to a large extent, to compensate the energy costs of the biogas compressor during the preliminary water absorption or the heat costs for the regeneration of the saturated amine solution. Thus, when compensating the power compressing biogas to 0.26 MPa, the specific energy consumption amounted to an average of 0.42 kWh/nm^3 of methane, which is 1.2 times higher than the energy consumption of traditional water absorption at P = 1.0 MPa. For amines, the desorber's reboiler energy consumption was an average of 0.29 kWh per nm^3 of methane, which is 20% less than water technology. Methane losses for both options of the water–amine absorption process with compensation of energy consumption are the same and range from 2.0 to 2.8%.

Figure 5.6 illustrates the specific energy consumption during the production of biomethane from the biogas flow 250 nm^3/h of different compositions using water, amine, and water–amine scrubbing processes. With a decrease in carbon dioxide concentration in biogas there is a sharp decrease in specific energy consumption in the amine absorption process, while water and water–amine absorption has an almost insignificant decrease in energy consumption for all compositions of biogas.

Energy consumption when using amine absorption of carbon dioxide from biogas is on average 2 times higher compared to water absorption, but the yield

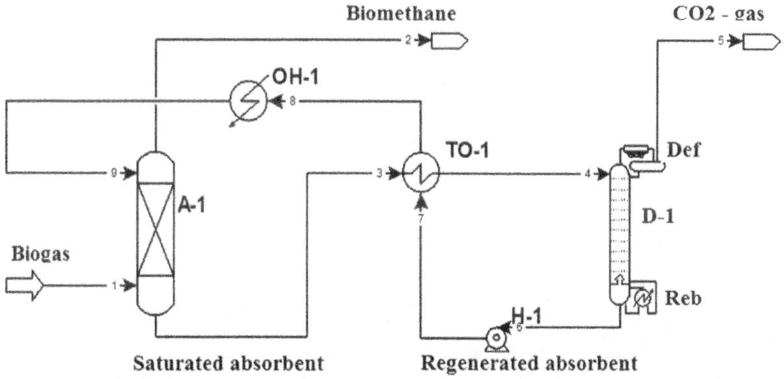

Figure 5.5 Technological flowchart of biogas amine scrubbing. A-1 – absorber; D-1 – desorber; Def – reflux condenser; Reb – reboiler; TO-1 – recuperative heat exchanger; OH-1 – heat exchanger (cooler); H-1 – pump.

Table 5.2 Results of simulation of water, amine, and water–amine of carbon dioxide absorption processes from 250 N m³/h biogas.

Indicators	CH_4/CO_2 Concentration in Biogas, % (vol.)		
	55/42	60/37	65/32
CH_4 content in biogas, N m³/h	137.5	150	162.5
Water pressure P_{abs} = 1.0 MPa (Var B)			
CH_4 content in biomethane, N m³/h	127	139	151
CH_4 content in degassing gas, % (vol.)	10.7	12.9	15.6
Energy consumption, kWh	47.7	47.7	47.7
Specific energy consumption, kWh/N m³_ CH_4	0.38	0.34	0.32
CH_4 losses, %	7.6	7.3	7.1
Amine pressure P_{abs} = 0.26MPa	137.5	150	162.5
CH_4 content in biomethane, N m³/h	119	110	101
Energy consumption, kWh	0.86	0.73	0.62
Specific energy consumption, kWh/N m³_CH_4 CH_4 losses, %	–	–	–
Water–amine P abs = 0.26 MPa (Var A)			
CH_4 content in biomethane, N m³/h	137.3	149.7	162
Energy consumption, kWh	72.8	76.3	79.5
Specific energy consumption, kWh/N m³_CH_4	0.53	0.51	0.48
CH_4 losses, %	0.15	0.2	0.3
Water–amine pressure P_{abs} = 0.26 MPa (Var B)*			
CH_4 content in biomethane, N m³/h	133.7	146.5	159.2
CH_4 content in degassing gas, % (vol.)	33.7	33.9	35.6
Energy consumption, kWh	57.8	62.5	67.5
Specific energy consumption, kWh/N m³_CH_4	0.43	0.43	0.42
CH_4 losses, %	2.8	2.3	2.0
Water–amine pressure P_{abs} = 0.26 MPa (Var B)**			
CH_4 content in biomethane, N m³/h	133.7	146.5	159.2
CH_4 content in degassing gas, % (vol.)	33.7	33.9	35.6
Energy consumption, kWh	36	42	48.5
Specific energy consumption, kWh/N m³_CH_4	0.27	0.29	0.3
CH_4 losses, %	2.8	2.4	2.0

*Combined water–amine absorption of CO_2 from biogas using potential energy of degassing gas ($CH_4 > 30\%$) to compensate the power consumption of the biogas compressor during the preliminary water absorption.**Combined water–amine absorption of CO_2 from biogas using potential energy of degassing gas ($CH_4 > 30\%$) to compensate the heat consumption for regeneration of saturated amine absorbent.*

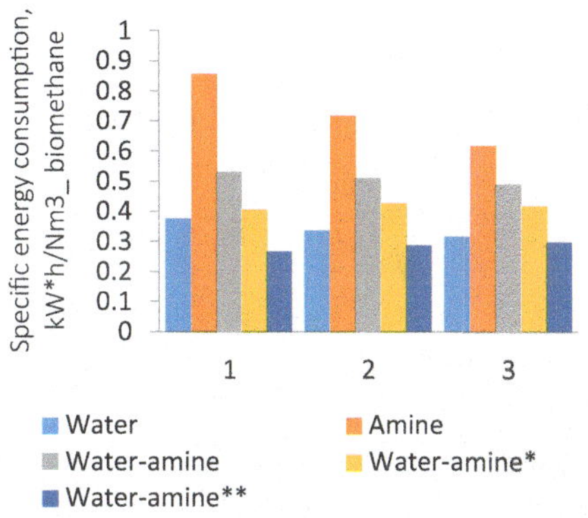

Figure 5.6 Specific energy costs of biomethane production from 250 nm³/h flow of different composition biogas using water, amine, and water–amine scrubbing processes. Content of CH_4/CO_2 in biogas, % (vol.): 1–55/42; 2–60/37; 3–65/32. *,** – as in Table 5.2.

of biomethane is 7.1–7.6% higher due to no methane losses, differs from the case of its dissolution in water.

5.3.4 Comparative results of computer simulation and laboratory testing

For the study of the biogas separation process with amines, the laboratory setup was developed at the gas technologies department of the Gas Institute. The scheme of laboratory setup includes a system for absorbent regeneration, which allows to heating of the saturated absorbent to the temperature of carbon dioxide intensive release. During these experiments, the process of carbon dioxide absorption and absorbent regeneration is studied (Figure 5.7). The specifications of the setup are the following:

Laboratory Setup Main Parameters	Specification
Diameter of absorber (internal)	76 mm
Diameter of desorber (internal)	129 mm
Temperature of the saturated absorbent at desorber input, not less	363 K (80°C)
Temperature of absorbent regeneration	373–393 K (100–120°C)
Maximum heat load of the desorber reboiler	2 kW
Biogas flow rate	0.5–2 m³/h
Absorbent flow rate	0.2–1.2 L/min (LPM)
Automatically controlled (stabilized) parameters	Regeneration temperature; absorbent flow

The basic chemical composition of biogas was taken as an average composition of landfill gas in Kamianets-Podilskyi (62% methane and 38% carbon dioxide).

Figure 5.7 Appearance of the laboratory setup.

Results of simulation and experimental results are shown in Table 5.3. Calculated and experimental data on the temperatures of the saturated absorbent at the outlet of the absorber and the regenerated absorbent at the outlet of the desorber are close enough and the difference does not exceed 5 K. The maximum deviation of the obtained concentration of biomethane was ± 2.8% at a pressure in the absorber Pabs = 0.12–0.2 MPa.

It is shown that for the amine absorption scheme of the laboratory installation of biomethane production without additional compression of biogas, the pressure in the absorber and desorber can be minimal, 0.12 and 0.11 MPa, respectively. At the same time, the estimated heat load on the reboiler does not exceed the accepted minimum - 1.5 kW.

The results of experiments showed that the software simulation model is correct and precise and can be used to simulate large-scale equipment installed on landfill power stations.

5.3.5 Enrichment of biogas with high air content

Increasing biogas flow to the engine or decreasing wells debit will decrease biomethane concentration. When the rate of biogas pumping from landfills increases, the concentration of methane can drop to 30% or less. At the same time, the critical methane content in fuel for gas piston engines is 28%. To

Table 5.3 Comparison of software simulation and experimental results.

Indicators	Pressure in the Absorber P_{abs}, MPa		
	0.12	0.16	0.2
	exp/calc		
Absorbent flow rate, kg/h	18/19	15/15	14/13
Regenerated absorbent temperature at the absorber entrance, °C	39/39	52/55	53/55
Saturated absorbent temperature at the absorber outlet, °C	47/52	76/71	76/73
Saturated absorbent temperature at the desorber entrance, °C	83/83	92/92	98/98
Regenerated absorbent temperature at the desorber exit, °C	97/100	113/112	117/118
Concentration in biomethane, % (vol.)*:			
CH_4	95.5/95.4	93.8/95.7	92.6/95.2
CO_2	3.3/4.3	5.5/4.0	6.6/4.6
Heat load of the desorber reboiler, kWh	1.5	2.0	2.0

**The concentration of CH_4 in biomethane is given taking into account drying to the temperature of the dew point – 8°C (at P = 3.92 MPa).*

solve the problem, biogas enrichment (through carbon dioxide extraction) with increased air content was considered. The results of calculations of the carbon dioxide extraction process from different compositions of biogas with increased air (nitrogen 20–40% by volume) are shown in Table 5.4. The concentration of methane increased by 1.2–1.4 times, and the heat consumption of 1 kg of carbon dioxide extraction was 1 kW for the wide range of biogas composition.

Table 5.4 Indicators of the carbon dioxide extraction process from different compositions biogas with increased air content.

Indicators	Composition of Biogas, % (vol.)		
	CH_4–50 CO_2–30 N_2–20	CH_4–45 CO_2–25 N_2–30	CH_4–40 CO_2–20 N_2–40
MDEAmod absorbent flow rate, kg/h	**2318**	**1733**	**1346**
Content in extracted gas, % (vol.):			
CH_4	68.3	57.3	47.8
CO_2	1.2	1.2	1.1
Output of gaseous CO_2, kg/h	143	118	92
Thermal load of desorber reboiler, kW	159	118	93
Specific power consumption, kW/kg_CO_2	1.1	1.0	1.0

5.4 CONCLUSIONS

Power station working at landfills is restricted by landfill gas composition. If methane content is below 30% (vol.) engine works are not stable. This low methane content is caused by not enough outside covering of landfill surface or its absence at all on acting landfills, and the case is typical for Ukraine.

It is possible to change the current state by enriching methane content through carbon dioxide extraction. Authors tried to choose a more effective method of carbon dioxide absorption taking into account energy consumption and methane losses with an outlet solution and the landfills available resources in Ukraine.

Figure 5.8 shows losses (%, vol.) of methane in the obtained biomethane in the considered processes. The most significant losses occur during water absorption, which is associated with increased solubility of biomethane in water. The amount of methane in biomethane in output flow practically corresponds to its input amount in biogas when applying the technology of amine and water–amine absorption.

For the water–amine process of carbon dioxide extraction from biogas, in which the potential of gas degassing is used, biomethane losses are 3 times lower than traditional water absorption.

The energy costs of combined water–amine absorption are commensurate with the energy costs when using water absorption due to:

- a significant reduction in biogas compression during the initial water absorption of carbon dioxide from biogas;
- reduction of heat costs for the regeneration of the saturated amine absorbent, as part of the carbon dioxide has already been previously removed from the biogas;

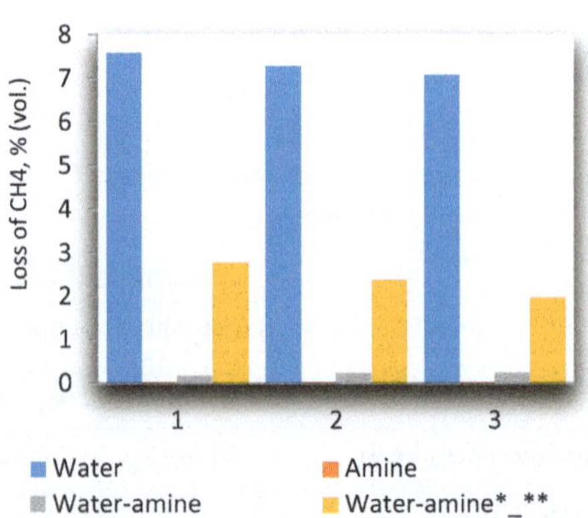

Figure 5.8 Methane losses during 250 nm³/h various composition biogas flow processing using water, amine, and water–amine absorption processes. Content of CH_4/CO_2 in biogas, % (vol.): 1–55/42; 2–60/37; 3–65/32. * and ** – combined water–amine absorption of carbon dioxide from biogas using potential energy of degassing gas (biomethane more than 30%).

- using aeration gas as fuel gas to compensate for power consumption of the biogas compressor during the preliminary water absorption of carbon dioxide or to compensate for the heat load of the reboiler of the amine desorber.

REFERENCES

Al-Masabi F. H. and Castier M. (2011). Simulation of carbon dioxide recovery from flue gases in aqueous 2-amino-2-methyl-1-propanol solutions. *International Journal of Greenhouse Gas Control*, **5**, 1478–1478, https://doi.org/10.1016/j.ijggc.2011.08.005

Anuchin K. M. and Miroshnichenko D. A. (2015). The possibility of using piperazine-activated methyldiethanolamine as an absorbent for deep gas purification from CO_2 based on calculations in the petro-SIM program. *Modern Technologies for Processing and Using Gas*, **1**, 9–16.

AR6 WGI Report. (2021). List of corrigenda to be implemented. The Intergovernmental Panel on Climate Change. Chapter 7. Supplementary Material. URL: https://www.ipcc.ch/report/ar6/wg1/downloads/report/IPCC_AR6_WGI_Chapter_07_Supplementary_Material.pdf

Capraa F., Fettarappab F., Maglia F., Gattib M. and Martellib E. (2018). Biogas upgrading by amine scrubbing: solvent comparison between MDEA and MDEA/MEA blend. *Energy Procedia*, **148**, 970–977, https://doi.org/10.1016/j.egypro.2018.08.065

Gang L., Liqing K. and Chao L. (2017) Absorption performance for CO_2 capture process using MDEA-AMP aqueous solution. *IOP Conference Series: Earth and Environmental Science*, **59**, 1–4, https://doi.org/10.1088/1755-1315/59/1/012011

Ivanov Y. V., Pyatnichko A. I., Zhuk G. V., Onopa L. R. and Soltanibereshne M. A. (2017). Extraction of carbon dioxide from gas mixtures with amines absorbing process. *Energy Procedia*, **128**, 240–247, https://doi.org/10.1016/j.egypro.2017.09.062

Ivanov Y. V., Zhuk G. V., Onopa L. R. and Krushnevych S. P. (2021). Comparative analysis of the effectiveness of water and water-amine absorption processes of CO_2 extraction from biogas. *Energy Technologies & Resource Saving*, **4**, 17–26 (Ukr.), https://doi.org/10.33070/etars.4.2021.02

Pyatnichko O., Zhuk H., Bannov V. and Kubenko S. (2013). System of landfill gas collection and utilisation. *International Journal of Energy for a Clean Environment*, **14**(2–3), 191–199, https://doi.org/10.1615/InterJEnerCleanEnv.2014006803

Saleem N. K., Sintayehu M. H., Zakaria M. and Azmi M. S. (2017). High pressure solubility of carbon dioxide (CO2) in aqueous solution of piperazine (PZ) activated N-methyldiethanolamine (MDEA) solvent for CO_2 capture. *AIP Conference Proceedings*, **1891**, 020081, https://doi.org/10.1063/1.5005414

Tashyrev O., Hovorukha V., Havryliuk O., Sioma I., Gladka G., Kalinichenko O., Włodarczyk P., Suszanowicz D., Zhuk H. and Ivanov Y. (2022). Spatial succession for degradation of solid multicomponent food waste and purification of toxic leachate with the obtaining of biohydrogen and biomethane. *Energies*, **15**, 911, https://doi.org/10.3390/en15030911

Zhuk H., Ivanov Y., Onopa L., Krushnevych S. and Soltanibereshne M. (2024). Effectiveness of Water-Amine Combined Process for CO2 Extraction from Biogas. *Environmental and Climate Technologies*, **28**(1), 135–148, https://doi.org/10.2478/rtuect-2024-0012

doi: 10.2166/9781789065138_0069

Chapter 6

Environmental conditions regarding CO_2 emissions in small- and medium-sized enterprises

M. Majka[1], B. Kuzio-Wasilewska[1], B. Woźniak[1] and M. Zabochnicka[2]

[1]IGO Sp. z o.o., Barbary 21, 40-053 Katowice, Poland
[2]Czestochowa University of Technology, Faculty of Infrastructure and Environment, J.H. Dąbrowskiego 69, 42-201 Czestochowa, Poland
**Corresponding authors: kuzio@odpady.biz.pl; majka@odpady.biz.pl; magdalena.zabochnicka@pcz.pl*

ABSTRACT

Entities emitting CO_2 are obliged to take appropriate steps to protect the environment. The Environmental Protection Law and the Greenhouse Gas and Other Substances Emissions Management System Act contain provisions and regulations regarding the above-mentioned compound. Owners of enterprises responsible for introducing CO_2 into the environment must: obtain an administrative decision regarding the use of the environment, prepare annual emission reports, pay environmental fees, and submit appropriate reports. Research conducted among domestic enterprises from the SME sector showed that: the vast majority of surveyed enterprises use fuel in the form of hard coal, eco-pea coal and eco-pea coal for heating and generating hot water. The second dominant heating source is natural gas. Other fuels used are fuel oil and liquefied petroleum gas. Wood is used least often for heating. Progressing technology means that entrepreneurs are increasingly equipping their companies with modern heating furnaces and more ecological fuels. More ecological types of fuel include natural gas or eco-pea coal. A thorough analysis of CO_2 emissions in the sector of small- and medium-sized enterprises has identified the industry that needs the most demanding changes. These are poultry producers. These entrepreneurs use significant amounts of coal. They often have archaic combustion stoves. In addition, they do not pay fees for releasing gases and dust into the air for heating poultry houses. They are also exempt from submitting information in this regard. Further research in this sector will allow for a deeper understanding of the problem of CO_2 emissions and how to reduce them.

Key words: Emissions, CO_2, environment, fees, gases

6.1 INTRODUCTION

Industries release many contaminants to the environment – liquid, solid and gaseous (Zabochnicka, 2022; Zabochnicka-Świątek, 2013). Industrial contaminants have to be managed according to the law (Bargieł and Zabochnicka-Świątek, 2018). Entities emitting CO_2 are obliged to take appropriate steps to protect the environment (Krzywonos *et al.*, 2014). The Environmental Protection Law and the Greenhouse Gas and Other Substances Emissions Management System Act contain provisions and regulations regarding the above-mentioned compound. Owners of enterprises responsible for introducing CO_2 into the environment must: obtain an administrative decision regarding the use of the environment, prepare annual emission reports, pay environmental fees, and submit appropriate reports. These obligations also apply to small- and medium-sized enterprises. For this purpose, a market survey of entrepreneurs was carried out as part of the Eureka project. The above study allowed for the determination of a number of emission conditions, that is: the type of fuels used, the types of industrial installations used and the amount of fuels used. This contributed to better understanding the topic of CO_2 emissions. The analysis of the aspects identified the companies that most require reducing exhaust emissions. Then, representatives were selected for testing using an innovative CO_2-binding sorbent.

6.1.1 CO_2 emissions according to the Environmental Protection Law

The conditions for using the environment by entrepreneurs in terms of CO_2 emissions are determined mainly by the Act of April 27, 2021, Environmental Protection Law (Journal of Laws of 2024, item 54, consolidated text). It is supplemented by the Act on the System for Managing Emissions of Greenhouse Gases and Other Substances (Journal of Laws of 2022, item 673). The Environmental Protection Law defines the rules for the use of the environment and its resources. They concern the introduction of substances or energy into the environment and the costs associated with the environment.

The induction of substances, for example CO_2, into the environment requires administrative decisions. These are: a permit for releasing gases and dust into the air or an integrated permit. Smaller installations only require notification.

Section IV of the Environmental Protection Law states: "*operation of installations causing the release of gases or dust into the air is permitted after obtaining a permit, if required*". This permit is issued at the request of the operator of the installation. An application for a permit should include, among others: the name and address of the operator of the installation, type of installation, type of equipment and technology, source and location of emissions and their size. Emissions from installations are measured, if required. At the same time, supporting actions are outlined to prevent or reduce emissions. The permit is issued for a specified period of time, but no longer than 10 years. The exception is the integrated permit. It is issued for an indefinite period. However, it may be issued for a specified period, at the user's request.

Integrated permits are required when an installation whose operation may cause a significant negative impact on environmental conditions. Integrated permits cover all environmental impacts specified in sector permits and their interconnections. The provisions included in the integrated permit depend on the nature and scale of the activities conducted using the installation. If an integrated permit is issued, a separate emission permit is not needed. The integrated permit contains information about the installation used. These details are provided in the Regulation of the Minister of the Environment of August 27, 2014 on types of installations that may cause significant pollution of individual natural elements or the environment as a whole (Journal of Laws of 2014, item 1169).

6.1.2 Conditions for reporting an installation

An installation that has a minor impact on the environment only requires notification.

The application should include:

- type and scope of activities conducted;
- volume of production or services provided;
- name of the operator of the installation, his residence or registered office address;
- address of the plant where the installation is operated;
- installation operation time (days of the week and hours);
- volume and type of emissions;
- description of the emission reduction methods used;
- information whether the form of emission reduction is consistent with applicable regulations.

Notification of installations is regulated in: the Regulation of the Minister of the Environment of July 2, 2010 on cases in which the release of gases or dust into the air from installations does not require a permit (Journal of Laws of 2010, No. 130, item 881) and announcement of the Minister of the Environment of July 22, 2019 on the publication of the uniform text of the regulation of the Minister of the Environment on the types of installations whose operation requires notification (Journal of Laws of 2019, item 1510).

The regulation specifies the types of installations that require notification. Energy installations with a nominal thermal power of up to 5 MW fired with hard coal and up to 10–15 MW with other fuels, and in the case of non-energy fuels up to 1 MW, do not require a permit, that is they must be reported.

According to the above regulation, the following installations are excluded from notification:

- energy installations–with a nominal thermal power of less than 1 MW,
- non-energy installations–with a nominal thermal power of up to 0.5 MW.

6.1.3 Environmental fees

Induction of gases and dust into the air is one of the types of use of the environment. It is subject to a fee. This also applies to CO_2 emissions.

The entrepreneur is obliged to determine the amount of fees for using the environment. Fees are payable to the Marshal's Office by March 31 of the following year. Payment is made together with a statement of the fees due. Fees for releasing gases and dust into the air are determined on the basis of the annual emissions included in the so-called KOBIZE report.

The exception are fees for the use of the environment, the annual amount of which for one type of use of the environment does not exceed PLN 800. If the annual fee for one type of environmental use does not exceed PLN 100, no lists of the scope of environmental use are submitted to voivodeship marshals. Fees are paid in forms regulated by the Regulation of the Minister of Climate of December 11, 2019 on lists containing information and data on the scope of use of the environment and the amount of fees due (Journal of Laws of 2019, item 2443). The amount of the fee for releasing gases and dust into the air is determined in accordance with current regulations. The applicable rates are included in the announcement of the Minister of Climate and Environment of October 7, 2022 on the rates of fees for the use of the environment for 2023 (Journal of Laws of 2022, item 1009).

In the absence of a proper permit for releasing gases or dust into the air, a fee increased by 500% is payable (Article 292 of the Environmental Protection Program). However, this does not apply to CO_2 emissions. There is no emission limit specified in the permit for this substance. There are no defined limit levels or values in the air.

Entities using the environment are excluded from financial liability for the use of the environment in terms of emissions of substances into the air when:

- rescue operations are being carried out,
- is a natural person who is not an entrepreneur and does not need to have a permit to release substances into the air,
- the amount charged to the account of a given marshal's office does not exceed PLN 800 for total air emissions, while submitting the list,
- has an installation covered by the greenhouse gas emission allowance trading system and which has been granted emission allowances. A fee is then required to be paid to the bank account of the National Fund for Environmental Protection and Water Management in an amount equal to the product of the number of emission allowances granted for the first year and the unit fee for introducing CO_2 into the air applicable in the previous year. In such a case, the organization does not incur any fees to the account of the Marshal's Office in this settlement period.

6.1.4 CO_2 emissions in the Act on the greenhouse gas and other substances emission management system

The second pillar of information related to CO_2 emissions is the Act on the system for managing emissions of greenhouse gases and other substances. It should be noted that the list of fees submitted to the Marshal Office should be prepared on the basis of a report to KOBiZE. Both documents should be consistent with each other. This act obliges entities using the environment to prepare and submit to KOBiZE a report on the previous calendar year. It is submitted by the end of February. It contains detailed information on the amount of greenhouse gases and other substances emitted into the air and on the activities conducted that result in emissions.

6.1.5 CO_2 emissions in Poland

According to data from the Central Statistical Office (2021), emissions of gaseous pollutants from plants with the highest emissions of air pollutants in the country (1841 plants) amounted to approximately 210 million tonnes. It increased compared to the previous year (2020) by approximately 13%. More than 99% of these emissions were carbon dioxide. The rest were: sulfur dioxide, carbon monoxide, and nitrogen oxides. The main sources of gaseous pollutant emissions were units producing and supplying electricity, gas, steam, hot water (over 70%) and industrial processing plants (28%) (GUS, 2021).

The analysis of air emissions from small- and medium-sized enterprises indicates that most of them are emissions from the combustion of fuels for energy purposes, and to a lesser extent from technological processes (Update of the National Air Protection Program until 2025, 2021).

6.1.6 Small- and medium-sized enterprises and CO_2 emissions

Research was carried out in small- and medium-sized companies characterized by high emissions of gases into the atmosphere resulting from combustion. The data showed that these companies usually have energy installations with a nominal capacity of up to 5 MW. Most often, these were installations with a capacity of up to 1.4 MW or smaller, fired with hard coal, natural gas, liquefied petroleum gas, fuel oil, or wood. Table 6.1 shows the list of energy carriers most frequently used in small- and medium-sized enterprises and the degree of their share (Kuzio *et al.*, 2022).

New boiler models dominate among the surveyed entrepreneurs. The fuels used are: natural gas, liquefied propane-butane gas, and eco-pea coal. About 37.5% of entrepreneurs use hard coal for energy purposes, 10% use eco-pea coal, 2.5% coal fines and 17.5% heating oil. Natural gas (20%) and liquefied petroleum gas (LPG) (10%) are used by 30% of entrepreneurs. The share of wood is 2.5% (Majka, 2022a; Majka, 2022b; Majka *et al.*, 2019; Potępa -Błędzińska, 2021).

Table 6.1 List of energy carriers most frequently used by the surveyed entrepreneurs.

Fuel Type	Unit	Number of old (S) and New (N) Boilers	Number of Entities	Fuel Share (%)
Hard coal	Mg	15 – S, 1 -N	16	37,5
Eco-pea coal	Mg	4 – N	4	10,0
Coal fines	Mg	1 – S	1	2,5
Heating oil	m^3	6-N, 1-S	7	17,5
Natural gas	m^3	8 -N	8	20,0
Wood	Mg	1 -S	1	2,5
Liquefied gas propane butane	m^3	4 -N	4	10,0
Total		**18 -S, 23-N**	**41**	**100**

Source: Own study according to market research.

6.1.7 CO_2 emission fees for small- and medium-sized enterprises

Users of the installation determine on their own the amount of fees due for the use of the environment.

The fees are paid to the account of the given marshal's office by March 31 of the following year, together with a list of the amounts due. As mentioned earlier, no fees are payable for the use of the environment if the annual amount for one type of environmental use does not exceed PLN 800 and if the annual fee for one type of environmental use does not exceed PLN 100.

An exception to the fees applies to farmers breeding poultry who are not covered by the integrated permit. Farmers do not pay fees for releasing gases and dust into the air for heating poultry houses. They also do not have to provide information in this regard. This results from the content of the regulation on cases in which the introduction of gases or dust into the air from installations does not require a permit (Journal of Laws of 2010, No. 130, item 881; Journal of Laws of 2024, item 54), which is confirmed by the interpretation of the Ministry from 2009.

6.2 SUMMARY

Research conducted among domestic enterprises from the SME sector showed that: the vast majority of surveyed enterprises use fuel in the form of hard coal, eco-pea coal, and eco-pea coal for heating and generating hot water.

The second dominant heating source is natural gas. It is used, for example, as a heating medium in bakery ovens or in a steam generator for the production of rubber bands for mining.

Other fuels used are fuel oil and liquefied petroleum gas. Wood is used least often for heating.

Progressing technology means that entrepreneurs are increasingly equipping their companies with modern heating furnaces and more ecological fuels. More ecological types of fuel include natural gas or eco-pea coal.

A thorough analysis of CO_2 emissions in the sector of small- and medium-sized enterprises has identified the industry that needs the most demanding changes. These are poultry producers. These entrepreneurs use significant amounts of coal. They often have archaic combustion stoves. In addition, they do not pay fees for releasing gases and dust into the air for heating poultry houses. They are also exempt from submitting information in this regard. Further research in this sector will allow for a deeper understanding of the problem of CO_2 emissions and how to reduce them.

ACKNOWLEDGMENTS

This work was supported in the framework of the EUREKA NETWORK project "Capture of carbon dioxide by innovative sorbent – InnoCO2sorbent" (EUREKA/InnoCO2Sorbent/2/2021), financed by the National Centre for Research and Development.

REFERENCES

Bargieł P. and Zabochnicka-Świątek M. (2018). Technologies of coke wastewater treatment in the frame of legislation in force. *Ochrona Środowiska i Zasobów Naturalnych*, **29**(1), 11–15, https://doi.org/10.2478/oszn-2018-0003

GUS Central Statistical Office, *Environmental Protection in 2021*. (In Polish)

Journal of Laws of 2024, item 54. Announcement of the Marshal of the Sejm of the Republic of Poland of 7 December 2023 on the announcement of the consolidated text of the Environmental Protection Law Act. (In Polish)

Journal of Laws of 2022, item 673. Announcement of the Marshal of the Sejm of the Republic of Poland of 3 March 2022 on the announcement of the consolidated text of the Act on the system for managing emissions of greenhouse gases and other substances. (In Polish)

Journal of Laws of 2022, item 1009. Announcement of the Minister of Climate and Environment of October 7, 2022, regarding the rates of fees for the use of the environment for 2023. (In Polish)

Journal of Laws of 2019, item 1510. Notice of the Minister of the Environment of 22 July 2019 on the publication of the consolidated text of the regulation of the Minister of the Environment on the types of installations whose operation requires notification. (In Polish)

Journal of Laws of 2019, item 2443. Regulation of the Minister of Climate of 11 December 2019 on lists containing information and data on the scope of use of the environment and the amount of fees due. (In Polish)

Journal of Laws of 2014, item 1169. Regulation of the Minister of the Environment of 27 August 2014 on the types of installations that may cause significant pollution of individual natural elements or the environment as a whole. (In Polish)

Journal of Laws of 2010, No. 130, item 881. Regulation of the Minister of the Environment of 2 July 2010 on cases in which the release of gases or dust into the air from installations does not require a permit. (In Polish)

Krzywonos M., Borowski PF, Kupczyk A. and Zabochnicka-Świątek M. (2014). Abatement of CO_2 emissions by using motor biofuels, *Przemysl Chemiczny*, **93**(7), 1124–1127.

Kuzio S., Kuzio-Wasilewska B. and Majka M. (2022). Market study on entities emitting CO_2, "Task 2 within the Eureka project". "Carbon dioxide capture by an innovative sorbent". IGO Sp. z o.o., Katowice, Poland. (In Polish)

Majka M. (2022a). Report for the site using the environment: "XTRIM" Zbigniew Michella General Partnership, constituting part of the report for the entity "XTRIM" Zbigniew Michella General Partnership. KOBiZE. (In Polish)

Majka M. (2022b). Report for the site using the environment: "ZELKOT - Brzezina, Urzynicok General Partnership", constituting part of the report for the entity Boiler Components Plant "ZELKOT" Alojzy Brzezina and Henryk Urzynicok General Partnership. KOBiZE. (In Polish)

Majka M., Potępa-Błędzińska Z., and Kuzio-Wasilewska B. (2019). Notification of an installation from which emissions do not require a permit, for Cobra Europe Sp. z o.o., based in Piekary Śląskie. Report by IGO Sp. z o.o., Katowice. (In Polish)

Potępa-Błędzińska Z. (2021). Report for the site using the environment: "Elektrocarbon – Galwan Sp. z o.o.", constituting part of the report for the entity "Elektrocarbon - Galwan Sp. z o.o." KOBiZE. (In Polish)

Update of the National Air Protection Program until 2025 (with a perspective until 2030 and 2040) (2021). Ministry of Climate and Environment, Warsaw. (In Polish)

Zabochnicka M. (2022). Industrial wastewater as a growth medium for microalgal biomass for a sustainable circular bioeconomy. *Applied Sciences*, **12**(20), 10299, https://doi.org/10.3390/app122010299

Zabochnicka-Świątek M. (2013). Utilization of *Chlorella vulgaris* and sediments after N-NH_4 removal containing clinoptilolite for sorption of heavy metals from wastewater. *Rocznik Ochrona Srodowiska/Annual Set the Environment Protection*, **15**(1), 324–347.

Part 2

doi: 10.2166/9781789065138_0079

Chapter 7

Plasma-assisted gasification of solid organic wastes

O. M. Dudnyk[1,2], V. A. Zhovtyansky[2] and M. V. Ostapchuk[2]*

[1]Thermal Energy Technology Institute of National Academy of Sciences of Ukraine, 19, Andriivska Str., Kyiv, Ukraine
[2]The Gas Institute of National Academy of Sciences of Ukraine, 39, Dehtyarivska Str., Kyiv, Ukraine
**Corresponding author: aldudnyk2018@gmail.com*

ABSTRACT

Plasma steam gasification of biomass (bagasse and fast-growing algae) and organic waste (solid domestic waste and sewage sludge) was studied at the plasma steam gasification plant. Dry synthesis gas with hydrogen content of more than 45.3 vol. % was obtained as a result of plasma steam gasification of all types of organic raw materials. Bio-char with a LHV of 27.6 MJ/kg were produced as a result of incomplete plasma steam gasification of bagasse.

The process of partial oxidation of char by air was used at plasma steam–air gasification plant to increase the rate of conversion of the obtained char and steam use degree. The gas with H_2/CO molar ratio of 1.4–1.5 at a molar ratio of steam/oxygen in the blast of 1.2–1.3 was obtained during plasma steam–air gasification of mixture of sewage sludge pellets and rubber crumbs of worn tires.

Analysis of the results of experimental studies of the conversion of solid organic wastes and biomass showed that the use of a steam plasma torch enables to reduce the impact of solid organic wastes (organic fermentation products – greenhouse gases CH_4 and CO_2) on the environment and to obtain additional valuable products – bio-char and hydrogen-enriched synthesis gas.

Keywords: Plasma, waste, hydrogen, steam, air

7.1 INTRODUCTION

Plasma is one of the techniques that enable achieving renewable energy from solid residues, contributing to landfill avoidance and resource reutilization in line with the circular economy principles and supporting United Nations

Sustainable Development Goals 7 (affordable and clean energy), 12 (responsible consumption and production), and 13 (climate action) (Oliveira *et al.*, 2022).

In the past few years, rising concerns *vis-à-vis* global climate change and clean energy demand have brought worldwide attention to developing the "biomass/organic waste-to-energy" concept as a zero-emission, environmentally friendly and sustainable pathway to simultaneously quench the global energy thirst and process diverse biomass/organic waste streams. Bioenergy with carbon capture and storage (BECCS) can be an influential technological route to curb climate change to a significant extent by preventing CO_2 discharge. One of the pathways to realize this strategy is *via in situ* CO_2-sorption coupled with a thermal plasma gasification process (Sikarwar *et al.*, 2022).

Thermal plasma technologies offer advantages of efficient waste treatment due to its high temperature and energy density, lower pollutant emissions, rapid start-up and shut-down, and smaller size of the installation (Cai & Du, 2021). Processing of solid organic waste in plasma reactors is used to produce gaseous fuel, an alternative to natural gas (Fabry *et al.*, 2013). When solid organic waste (i.e., carbon-containing) is converted into an alternative gas fuel and can be used for energy self-sufficiency of the gasification plant, for example, using gas piston power plants (Diaz *et al.*, 2014) or as a separate fuel, increasing the overall energy efficiency of the process.

Plasma torches are introduced to generate electricity at centralized and autonomous power plants. In a study by Pan *et al.* (2022), a hybrid design that combines waste gasification and coal-fired power generation has been proposed for improving the waste-to-energy process. In the integrated scheme, municipal solid waste is fed into the plasma gasifier and converted into syngas, which is precooled by the feedwater of the coal power plant and then conveyed directly into the coal-fired boiler for combustion. The research conducted by University of Delft (the Netherlands) and Sunfire GmbH (Germany) developed a new process for production of electricity with the use of plasma conversion of solid organic waste and solid oxide fuel cell electrochemical generator (Fernandes *et al.*, 2018).

Plasma torches are used also as the main equipment for the direct conversion of organic raw materials (Zhovtyansky & Valinčius, 2018), as well as auxiliary (assisted) equipment for the conversion of solid organic waste. In addition, they are used to heat oxidizers, increase gas yield and produce vitrified slag (Zhang *et al.*, 2012). Reducing the content of macromolecular compounds (tar) in synthesis gas and obtaining synthesis gas from industrial gases is ensured using plasma torches. Studies by Materazzi and Taylor (2019) and Zhovtyansky *et al.* (2013a) show the results of plasma conversion of synthesis gas obtained after gasification of wood in the moving bed and RDF in the fluidized bed. The obtained synthesis gas can be used for the production of hydrogen (Dudnyk & Sokolovska, 2010) and motor fuel. Tamošiūnas *et al.* (2014) used an experimental plasma-chemical reactor with a steam-water plasma torch for propane reforming in an equilibrium plasma without the use of catalysts to obtain synthesis gas. Steam plasma torches can be used for catalytic conversion of solid organic waste into hydrogen (Singh *et al.*, 2019).

The use of steam in plasma torches is a key factor in obtaining hydrogen-enriched gas (Favas *et al.*, 2017). Water could be achieved from reconditioning, as reused or recycled industrial process water that helps to decrease water footprint.

In addition to the fact that water as a gasifying agent in plasma-steam conversion technologies allows obtaining gasification products with maximal heat value, it also optimally solves the problem of reducing nitrogen oxides from getting into these products – an inevitable attribute of plasma-air technologies, as well as dioxins and furans in the case of processing chlorine-containing raw materials (Zhovtyansky & Ostapchuk, 2022).

The purpose of the research (Lee *et al.*, 2014) was to study the gasification of coal for the production of synthesis gas using a system with an arc steam plasma torch. The influence of coal feed rate on chemical efficiency, degree of carbon conversion, and hydrogen yield was evaluated by the yield and composition of synthesis gas from two types of coal: I – low-grade coal with a high moisture content and II – high-grade coal (with a high lower heating value (LHV) and a low moisture content). The ratio of carbon to oxygen varied from 0.3 to 0.8 for coal I and from 0.5 to 1.2 for coal II, respectively. More synthesis gas was obtained from coal II than from coal I. In the case of using coal II, the degree of conversion of coal into synthesis gas, the cold gas efficiency and the heating value of dry synthesis gas were 89.0%, 58.2% and 11.4 MJ/nm^3, respectively.

The paper (Uhm *et al.*, 2014) show the results of gasification of Indonesian high-ash lignite using two microwave steam plasma torches for heating the reaction chamber of a vortex-type gasifier. As a result, hydrogen-rich synthesis gas was obtained. The article states that in case of additional heating of synthesis gas due to partial oxidation of coal, the temperature inside the gasifier can reach 1700°C.

The paper (Zhovtianskyi *et al.*, 2011) shows description of experimental plasma-steam equipment for producing gas with high content of hydrogen from solid fuels and results of research of plasma-steam gasification of Ukrainian coals. The high-ash anthracite of Donetsk coal deposit and brown coal of Kirovograd region were used for research. During the study of plasma-steam gasification of anthracite, the composition of dry gas was, vol. %: H_2 – 61.6–62.3, CO – 11.53–13.5, CO_2 – 24.5–24.8, CH_4 – 0.3–0.4. During the study of plasma-steam gasification of brown coal, the composition of dry gas was, vol. %: H_2 – 61.4–66.1, CO – 1.8–10.6, CO_2 – 26.7–32.11, CH_4 – 0.01–1.33.

A process of high-temperature decomposition of lignite was studied on the plasma gasification reactor PLASGAS, where water-stabilized plasma torch was used as a source of high-enthalpy plasma (Serov *et al.*, 2019). The plasma torch capacity was 120 kW and allowed heating of the reactor to more than 1000°C. The material feeding rate in the gasification reactor was 30 or 60 kg per hour that is comparable to a small industrial production process. The ratio H_2/CO in obtained gas was in the range of 1.5–2.5 depending on the experimental conditions.

Both medical waste (Chernets *et al.*, 2008; Zhovtyansky *et al.*, 2013b) and sewage sludge (Petrov & Zhovtyansky, 2019; Zhovtyansky *et al.*, 2018) were converted using plasma-steam technologies. In both cases, the obtained syngas of high quality was obtained in steam plasma with a high enthalpy, which was generated in a steam-water DC plasma torch (Petrov & Zhovtyansky, 2019). As energy for the process is supplied by plasma and chemical reactions in the reaction products are not primary source of energy, the process can be applied for wide choice of organic materials and biomass. The process acts also as energy storage–electrical energy is transferred to plasma energy and then stored in produced syngas.

Petrov and Zhovtyansky (2019) considered, in addition to the processes of actual waste gasification in the solid phase using dense thermal plasma, the processes of conversion of carbon-containing impurities–pollution in water. In this case, it is most appropriate to use a non-stationary non-equilibrium plasma, in which kinetic plasma models are used to describe the processes (Arsentiev *et al.*, 2014; Zhovtyansky & Anisimova, 2014).

The aim of this work is to describe the results of research of solid organic waste conversion, carried out at equipment for plasma-steam gasification of solid organic material and the experimental plant for plasma steam-air gasification of solid organic waste with downdraft gasifier.

7.2 METHODS

The equipment for plasma-steam gasification of solid organic material was used for research of conversion solid household waste, sewage sludge, bagasse, fast-growing algae, and char from corncob waste.

The data of steam consumption during the process of solid organic material gasification, dynamics of steam gasification, composition of the obtained dry synthesis gas with high content of hydrogen, degree of carbon conversion during the process of experimental studies, and others were obtained at the equipment.

The equipment for plasma-steam gasification of solid organic material consists of plasma torch Multyplaz-3500 (1), high-temperature chamber of solid organic material gasification (2), cooler of obtained synthesis gas (3), tank for condensate (4), filter for residual moisture removal (5), flow meter (6), synthesis gas burner (7), chromatograph (8), and thermocouples (9) (Figure 7.1).

For plasma torch and synthesis gas cooler operation, an independent system for cooling the plasma torch nozzle and synthesis gas cooler with process water (input 14, output 15) and producing hot water is created and started in operation.

High-temperature chamber of solid fuel gasification consists of ceramic cartridge case (CCC) (11) for solid organic material loading (the reactor that can be loaded with fuel again after the plasma torch had been disconnected and cooled), layer of lightweight thermal insulation (10), and high-temperature filter at the outlet of gasification chamber (13). The cartridge is closed with two metal meshes (12). The layer of lightweight insulation is a ring set in a metal case made of heat-resistant steel. The high-temperature gasification chamber casing consists of two parts connected by flange.

Three thermocouples (9) are used in the equipment to measure temperatures. Thermocouples are mounted on the CCC outer wall for 30 mm from the

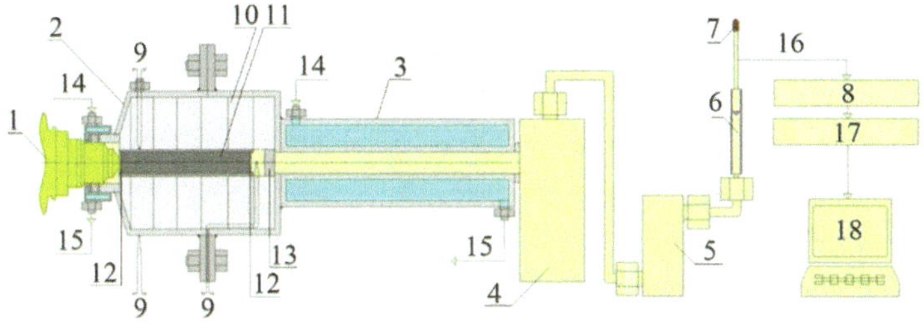

Figure 7.1 Equipment for plasma-steam gasification of solid organic material: 1 – plasma torch, 2 – high-temperature chamber of solid organic material gasification, 3 – synthesis gas cooler, 4 – tank for condensate, 5 – filter for residual moisture removal, 6 – flow meter, 7 – synthesis gas burner, 8 – chromatograph, 9 – thermocouple, 10 – thermal insulation, 11 – ceramic cartridge case with solid organic material, 12 – metal mesh, 13 – hot filter, 14 – input of cooling water, 15 – output of cooling water, 16 – sampling, 17 – interface, 18 – computer.

nozzle of plasma torch, on gasification chamber casing and before the high-temperature filter. Length of CCC is 140 mm, and wall thickness of ceramic tube is 3 mm. External diameter of high-temperature gasification chamber is 82 mm (wall thickness is 3 mm). Synthesis gas cooler is a heat-resistant metal pipe with external cooling by water. After cooling the synthesis gas, condensate is collected in the tank for condensate (4). Sampling gas (16) for analysis is conducted before the burner (16). Composition of obtained dry gas is measured by chromatograph (8). The gas composition data is transmitted through the interface (17) into the computer (18).

The test procedures are as follows. Sample of initial organic particles of given sizes was weighed and loaded into the CCC. The entrance and exit of the CCC were covered with metal meshes. High-temperature flow from plasma torch was directed through a hole in the middle of a metal mesh into the source fuel bed. During the experiment, CCC temperature, power capacity of plasma torch, consumption of the obtained gas, and its content (before the burner) were measured. After the gas had been released, the plasma torch was turned off, disconnected from the high-temperature chamber, cooled, and filled with water; the vent of the high-temperature chamber had closed with high-temperature wadding. After the CCC temperature had reduced to 50°C, the vent of the high-temperature chamber was opened and the CCC was removed from the high-temperature chamber. Residues of solid fuel conversion were removed from the CCC and fed to evaporation and drying. Dry residues of solid fuel conversion were weighted. The converted mass was determined by difference of sample weights before and after conversion. The reaction rate of the sample was determined by weight difference and time of reaction. The degree of carbon conversion was determined taking the analysis of the initial fuel and difference of sample weight (before and after conversion) into account.

The experimental plant for plasma steam-air gasification of solid organic waste with downdraft plasma steam-air gasifier was created to increase

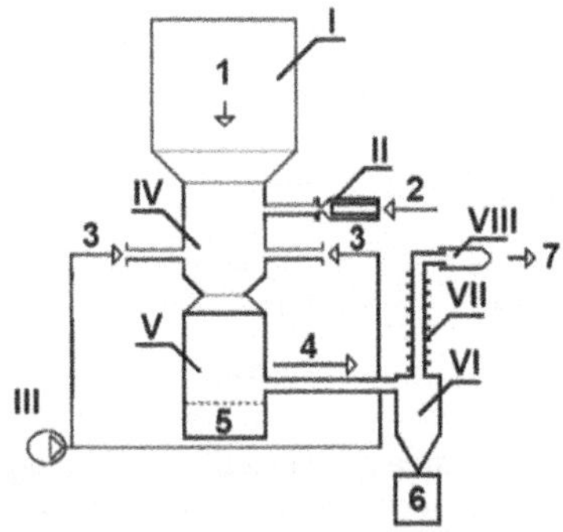

Figure 7.2 Experimental plant for plasma steam–air gasification of solid organic waste: I – solid organic waste hopper, II – steam plasma torch, III – compressor, IV – pyrolysis and partial oxidation zones of the gasifier, V – gasification zone of the gasifier, VI – cyclone with fly ash hopper, VII – cooler, VIII – ejection burner; 1 – solid organic waste, 2 – water, 3 – air, 4 – produced gas, 5 – bottom ash, 6 – fly ash, 7 – combustion products.

the productivity of the plasma-steam equipment using atmospheric oxygen (Zhovtyansky *et al.*, 2023, Dudnyk *et al.*, 2025). This plant was used for research of gasification of sewage sludge of aeration station of Kyiv and plasma steam-air gasification of mixture of pellet from sewage sludge of aeration station of Ivano-Frankivsk and rubber crumbs of worn tires (Figure 7.2).

The following equipment was used for the research: solid organic waste hopper (I), downdraft gasifier (IV–V), steam plasma torch Multiplaz 3500 (II), air compressor VK-50 (III), hot cyclone with fly ash hopper (VI), cooler of the obtained gas (VII) and ejection burner (VIII). Downdraft gasifier consists of pyrolysis, partial oxidation (IV) and gasification (V) zones. Three thermocouples were used to measure temperatures of the outer walls of the zones of pyrolysis, partial oxidation and gasification. Internal diameter of pyrolysis and gasification zones is 79 mm (wall thickness is 5 mm). Thermal insulation of gasifier and hot cyclone was used for decreasing heat losses and increasing efficiency of waste conversion.

The test procedures are as follows. The gasifier was started using a plasma torch and charcoal. Charcoal with a particle size of 10–20 mm was loaded into the gasifier and hopper. The hopper was closed. The plasma torch turned on. After 1 min, the plasma torch was turned off, and air was supplied to the gasifier by the compressor. After warming up the gasifier for 100 min, the air supply was stopped; the bunker was opened and loaded with solid waste. The particle size of solid waste was up to 10 mm. The bunker was closing; plasma torch and compressor were turned on.

Gas sampling was carried after cooler (VII). The composition of the dry obtained gas was determined by the chromatograph. Air flow rate is determined by flow meter. The water consumption in the plasma torch was determined depending on the electric capacity of the plasma torch (Zhovtianskyi *et al.*, 2011). As a result of the calculation of the mass and heat

Table 7.1 Compositions of the waste after drying (wt. %).

Waste	Moisture	Ash	Volatile Matter	Fixed Carbon
	wt. %			
Solid household waste	0.5	58.6	30.7	10.2
Sewage sludge	9.5	43.8	38.0	8.7
Bagasse	7.0	2.1	78.1	12.8
Fast-growing algae	5.3	22.5	59.9	12.3
Char from corncob waste	10.1	4.3	28.3	57.3
Sewage sludge pellets	6.5	60.6	26.8	6.1
Rubber crumbs of worn tires	1.4	23.3	57.4	17.9
Mixture of sewage sludge pellets and rubber crumbs of worn tires (50/50 wt./wt.)	3.9	42.0	42.1	12.0

balances of the experimental plant, the solid waste consumption, efficiency, and yield of hydrogen were determined.

The compositions of the wastes used for research after drying are presented in Table 7.1.

7.3 RESULTS AND DISCUSSION

7.3.1 Equipment for plasma-steam gasification

7.3.1.1 Solid household waste conversion

Solid household waste (SHW) was dried and shredded. Size of particles of SHW was 0.05–1.6 mm. 15 g of SHW was used for conversion. Figure 7.3 shows the results of plasma-steam gasification of SHW: electric capacity of the plasma torch, water consumption in the plasma torch, yield of dry obtained gas, composition of dry obtained gas, and carbon conversion degree.

The electric capacity of the plasma torch during the conversion of SHW was 440–650 W. The water consumption in the plasma torch was 1.4–1.9 g/min. The process of steam conversion of SHW has two main stages: (1) conversion of volatile matter (first 5 min); (2) gasification of fixed carbon (Figure 7.3c).

In the process of steam conversion of SHW, the composition of the obtained gas was, vol. %: H_2 – 45.3–64.4; CO – 2.1–14.2; CO_2 – 32.8–39.6; CH_4 – 0.7–1.5. Formula for steam conversion of SHW:

$$C_{1.000}H_{0.82}O_{0.593} + 1.108\ H_2O + 0.305\ CaCO_3$$
$$= 1.465\ H_2 + 0.246\ CO + 0.027\ CH_4 + 1.032\ CO_2 + 0.305\ CaO$$

The composition of all gas obtained during the plasma steam conversion of SHW, vol. %: H_2 – 52.9, CO – 8.9, CO_2 – 37.3, CH_4 – 0.9.

The degree of carbon conversion of solid household waste after the research study was 90.9%.

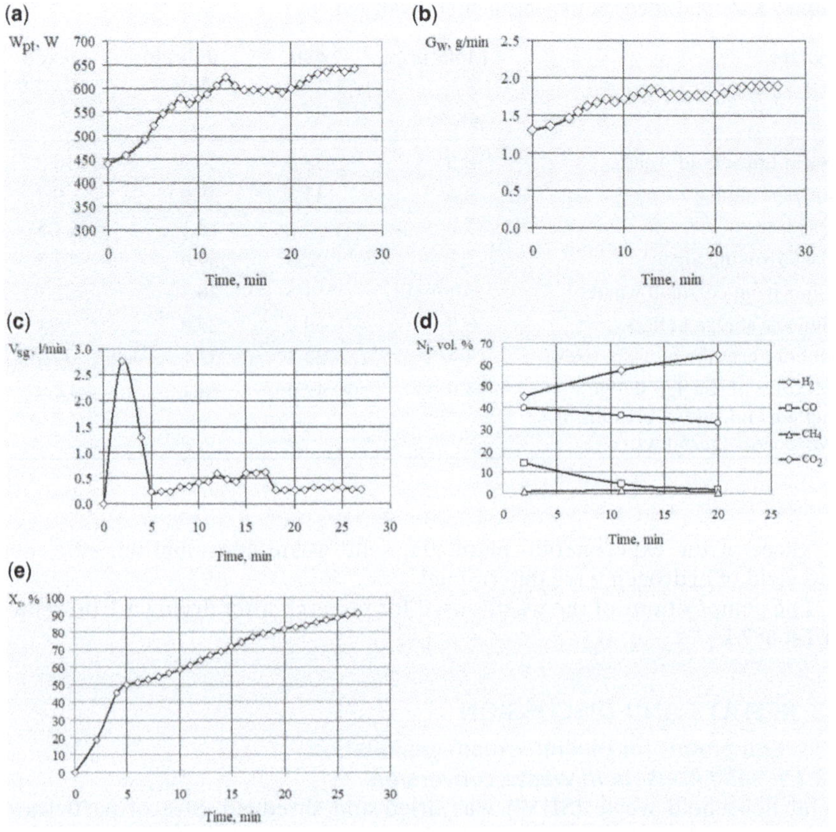

Figure 7.3 Results of plasma-steam gasification of solid household waste: electric capacity of the plasma torch (a), water consumption in the plasma torch (b), yield of dry obtained gas (c), composition of dry obtained gas (d) and carbon conversion degree (e).

7.3.1.2 Sewage sludge conversion

A measure of 7 g of dry sewage sludge (SS) was used for plasma-steam gasification. Figure 7.4 shows the results of plasma-steam gasification of SS: electric capacity of the plasma torch, water consumption in the plasma torch, yield of dry obtained gas, composition of dry obtained gas and carbon conversion degree.

The electrical capacity of the plasma torch was 500–520 W. The water consumption in the plasma torch is 1.5 g/min. 14 g of water was used in the plasma torch. The composition of the obtained gas was, vol. %: H_2 – 60.6–79.7, CO – 1.1–6.1, CO_2 – 19.2–32.6, CH_4 – 0.1–0.7. The flow rate of dry obtained gas varied from 0.26 to 1.12 L/min. The composition of the all produced gas was vol. %: H_2 – 71.8, CO – 3.1, CO_2 – 24.7, CH_4 – 0.4. 4.7 liters of dry synthesis

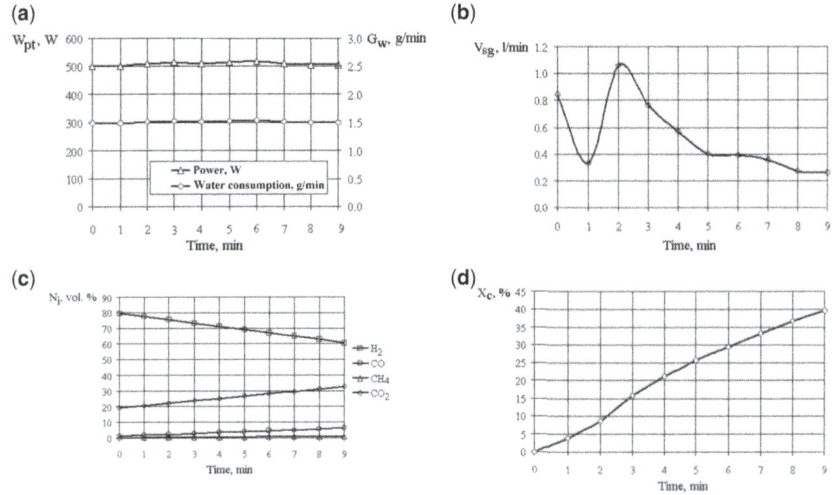

Figure 7.4 Results of plasma-steam gasification of sewage sludge: electric capacity of the plasma torch and water consumption in the plasma torch (a), yield of dry obtained gas (b), composition of dry obtained gas (c) and carbon conversion degree (d).

gas was obtained. LHV of all obtained dry gas was 8.3 MJ/Nm^3. The degree of carbon conversion of solid household waste after the research was 39.8%.

7.3.1.3 Bagasse conversion

A weight of 3 g of bagasse (solid waste of sugar production from sugar cane) was used for plasma-steam gasification. Figure 7.5 shows the results of plasma-steam gasification of bagasse: electric capacity of the plasma torch, water consumption in the plasma torch, yield of dry obtained gas, composition of dry obtained gas and carbon conversion degree.

The electric capacity of the plasma torch during bagasse conversion was 300–350 W, the water consumption in the plasma torch was 0.9–1.1 g/min. At the outlet of the equipment, dry synthesis gas composition was, vol. %: H_2 – 57.8–61.3, CO – 17.2–18.3, CO_2 – 19.2–23.7, CH_4 – 1.2–1.3. The degree of bagasse carbon conversion after the end of the experiment was 53.4%. In addition, 3.14 liters of dry synthesis gas and 0.77 g of charcoal were obtained. The composition of the total produced gas, vol. %: H_2 – 60.8, CO – 18.1, CO_2 – 19.9, CH_4 – 1.2. The composition of obtained charcoal was, wt. %: C – 77.1, H – 2.6, O – 10.4, ash content – 8.2, moisture content – 1.7. Lower heating value (LHV) of the charcoal was 27.6 MJ/kg.

7.3.1.4 Fast-growing algae conversion

Fast-growing algae (FGA) was a cultivated mixture of unicellular algae from the Dnipro River, Ukraine. The main component in algae was *Chlorella* sp. FGA is a third-generation biofuel and is considered in the world as a promising

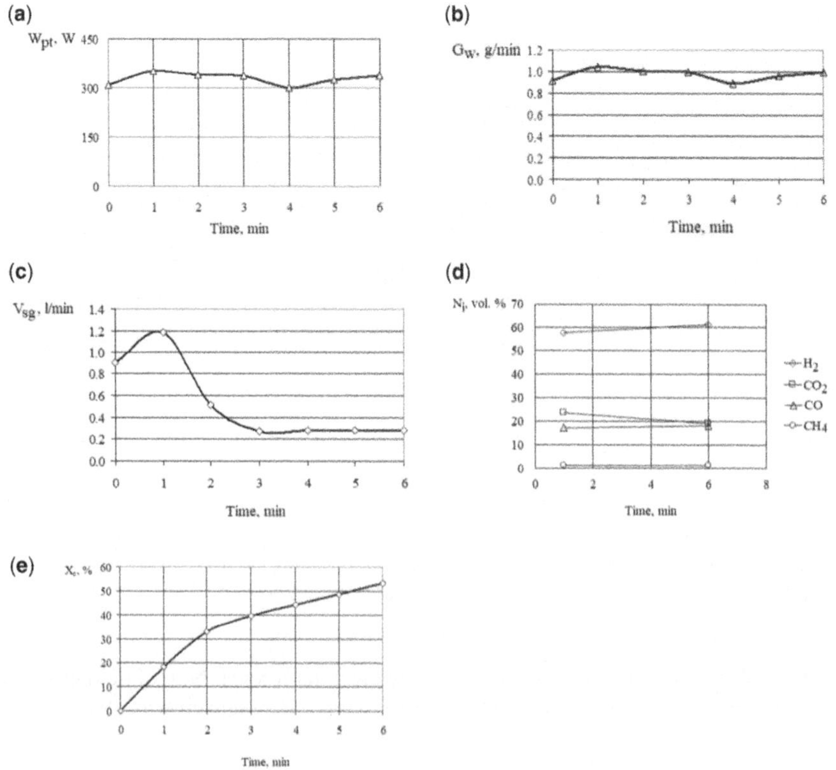

Figure 7.5 Results of plasma-steam gasification of bagasse: electric capacity of the plasma torch (a), water consumption in the plasma torch (b), yield of dry obtained gas (c), composition of dry obtained gas (d) and carbon conversion degree (e).

alternative solid fuel (the production of which reduces the carbon dioxide content in air). For plasma-steam gasification, algae were used after preliminary drying and grinding. The particle size of algae was 0.16–0.69 mm. Their specific feature of this dry feed stock was a significant ash content (more than 20%).

A measure of 25 g of algae after drying and grinding were loaded into a ceramic cartridge case before research. Bulk density of algae was 0.71 g/cm^3. Figure 7.6 shows the results of plasma-steam gasification of FGA: electric capacity of the plasma torch, water consumption in the plasma torch, yield of dry obtained gas, composition of dry obtained gas and carbon conversion degree.

The electric capacity of the plasma torch during the experiments was 990–1160 W. The water consumption in the plasma torch was 2.9–3.2 g/min. The maximum temperature of the outer wall of the ceramic cartridge case at a distance of 30 mm from the plasma torch nozzle was 1520°C.

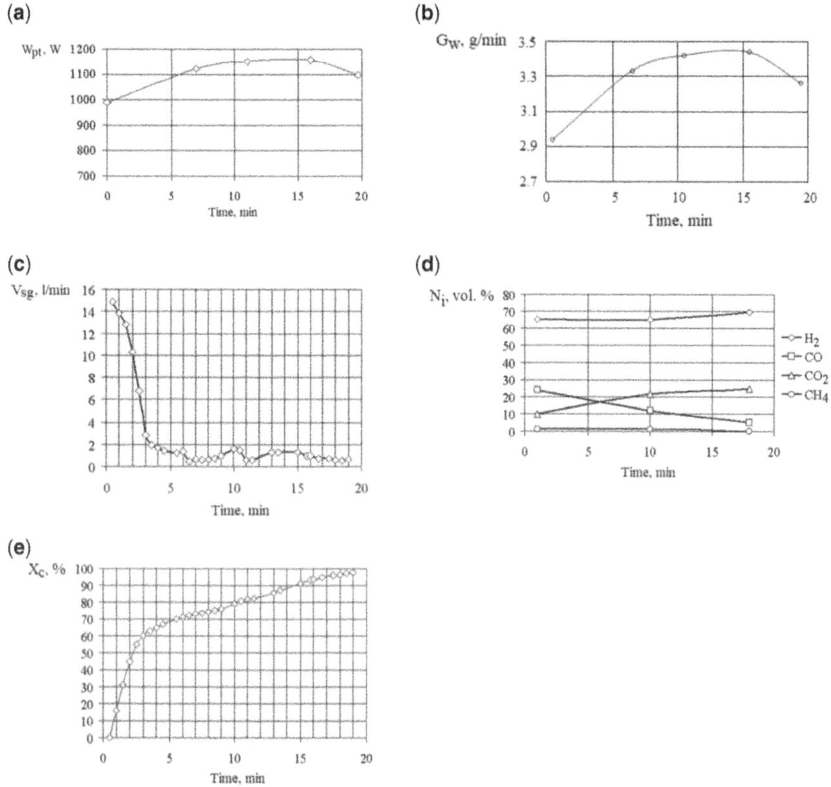

Figure 7.6 Results of plasma-steam gasification of fast-growing algae: electric capacity of the plasma torch (a), water consumption in the plasma torch (b), yield of dry obtained gas (c), composition of dry obtained gas (d) and carbon conversion degree (e).

The composition of the obtained gas was, vol. %: H_2 – 65.1–69.8, CO – 5.2–24.0, CO_2 – 9.8–24.9, CH_4 – 0.1–1.1. Mild thermal capacity of equipment for plasma-steam gasification at yield of obtaining gas in time of volatile matter conversion was 2.05 kW. Thermal capacity of equipment at yield of obtaining gas in time of fixed carbon conversion was up to 59 W.

Formula for steam conversion of FGA:

$$C_{1.00}H_{2.33}O_{0.61} + 0.7841\,H_2O = 0.548\,CO + 0.421\,CO_2 + 1.890\,H_2 + 0.030\,CH_4$$

The composition of all dry gas obtained from 25 g of algae is as follows, vol. %: H_2 – 65.4, CO – 19.0, CO_2 – 14.6, CH_4 – 1.0. Yield of all dry gas was 46.6 liters. LHV of all dry gas was 9.82 MJ/Nm^3.

At a consumption of 0.73 kWh of electricity and 60 g of water in the plasma torch, the degree of carbon conversion of fast-growing algae in a ceramic cartridge case was 97.6%.

7.3.2 Conversion of char coal from corncob waste

Waste of corncob (with ash content 1.2%) was crushed to a size of 0.315–10.0 mm. The obtained waste samples were carbonized at temperatures up to 812°C. As a result of carbonization, the LHV of solid fuel increased by 59% and its bulk density increased by 31%. The obtained charcoal (2.8 g) with a LHV of 23.5 MJ/kg, bulk density of 0.14 g/cm³ and ash content of 4.25% was gasified in the equipment for plasma-steam gasification.

Figure 7.7 shows the results of plasma-steam gasification of charcoal from corncob waste: electric capacity of the plasma torch, water consumption in the plasma torch, yield of dry obtained gas, composition of dry obtained gas and carbon conversion degree.

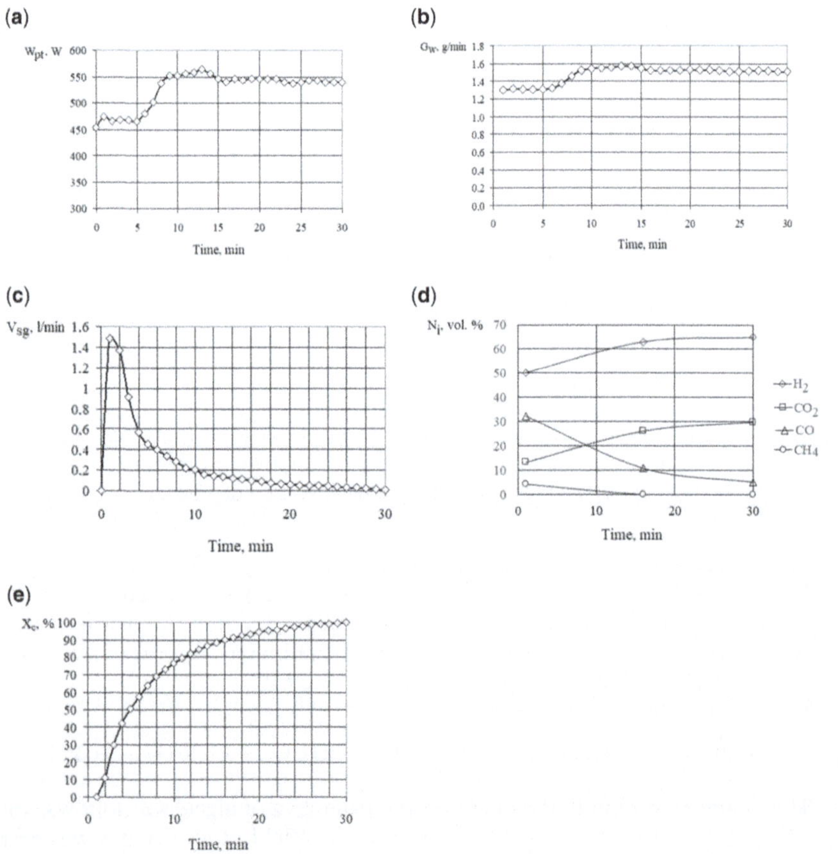

Figure 7.7 Results of plasma-steam gasification of charcoal from corncob waste: electric capacity of the plasma torch (a), water consumption in the plasma torch (b), yield of dry obtained gas (c), composition of dry obtained gas (d) and carbon conversion degree (e).

The electric capacity of the plasma torch was 453–564 W, the water consumption in the plasma torch was 1.3–1.6 g/min. The composition of the obtained gas was, vol. %: H_2 – 50.2–64.9, CO – 5.2–32.3, CO_2 – 3.2–29.9, CH_4 – 0.03–4.3.

During the 30 min of the experiment, a degree of conversion of fixed carbon of 99.8% was achieved, 0.264 kWh of electricity and 44.4 g of water were consumed.

7.3.3 The experimental plant for plasma steam-air gasification

7.3.3.1 Sewage sludge conversion

Table 7.2 shows the results of plasma steam-air gasification of sewage sludge.

Thermal capacity of experimental plant for plasma steam-air gasification was up to 3.6 kW by sewage sludge consumption and up to 1.83 kW at yield of dry obtained gas.

During plasma steam-air gasification, consumption of sewage sludge was from 1.3 to 2.1 kg/h for the electric capacity of the plasma torch 0.9–1.2 kW and the thermal capacity of the gasifier in relation to sewage sludge consumption from 2.1 to 3.6 kW. The yield of dry gas was from 1.4 to 2.6 Nm^3/h. Composition of dry gas obtained during plasma steam-air gasification, vol. %: H_2 – 11.5–20.7, CO – 6.0–10.7, CO_2 – 18.5–21.1, CH_4 – 0.2–1.5, N_2 – 48.6–61.7. Molar ratio of H_2/CO – 1.2–2.7. The yield of hydrogen was from 0.46 to 0.76 Nm^3/kg of dry ash free (DAF) sewage sludge.

7.3.3.2 Conversion of mixture of sewage sludge pellets and rubber crumbs of worn tires (50/50 wt./wt.)

Table 7.3 shows the results of plasma steam-air gasification of mixture of sewage sludge pellets and rubber crumbs of worn tires (50/50 wt./wt.).

Maximal thermal capacity of experimental plant for plasma steam-air gasification was 1.51 kW by mixture consumption and 0.73 kW at yield of dry obtained gas. The consumption of mixture of sewage sludge pellets and crumb from rubber tires was from 0.29 to 0.31 kg/h, the electric capacity of the plasma torch was 1.3–1.4 kW, and the thermal capacity of the gasifier in relation to the consumption of mixture was 1.4–1.5 kW. The yield of dry gas was from 0.95 to 0.98 Nm^3/h. The composition of the dry gas obtained during the plasma steam–air gasification of mixture, vol. %: H_2 – 11.0–12.7, CO – 8.1–8.8, CO_2 – 15.6–16.1, CH_4 – 0.6–0.7, N_2 – 62.4–64.1. Molar ratio of H_2/CO – 1.4–1.5. The yield of hydrogen was 0.65 Nm^3/kg of DAF waste mixture.

7.4 CONCLUSIONS

Different types of solid organic waste at the experimental installation for plasma-steam gasification of solid fuels were converted into hydrogen-enriched synthesis gas. Industrial water used for such processes could be recycled, reused, or reconditioned lowering the water footprint.

Activated by steam charcoal with a LHV of 27.6 MJ/kg and dry synthesis gas with hydrogen content of 60.8 vol. % were obtained as a result of not full plasma-steam gasification of bagasse at low electric capacity (300–350 W) during 6 min of conversion.

Table 7.2 Results of plasma steam-air gasification of sewage sludge.

Parameter	Value						Unit
Consumption of wet sewage sludge (moisture content is 42.3 wt. %)	1.26	1.65	1.51	1.79	2.12	1.92	kg/h
Thermal capacity by sewage sludge consumption	2.13	2.80	2.55	3.04	3.59	3.25	kW
Electric capacity of steam plasma torch	1.16	1.16	0.94	0.96	1.23	1.18	kW
Water consumption in plasma torch	2.49	2.49	2.09	2.17	2.61	2.52	g/min
Air consumption	0.84	1.22	1.18	1.51	1.89	1.89	Nm^3/h
Oxygen consumption	0.18	0.26	0.25	0.32	0.40	0.40	Nm^3/h
Steam consumption	0.15	0.15	0.12	0.13	0.16	0.15	kg/h
Steam/oxygen ratio	1.05	0.73	0.64	0.50	0.48	0.48	mol/mol
Temperature of the outer walls of the partial oxidation zone	481	562	610	611	669	701	°C
Yield of dry obtained gas	1.37	1.82	1.64	2.13	2.63	2.41	Nm^3/h
Composition of dry obtained gas							
H_2	20.65	16.78	12.44	15.39	16.16	11.53	vol. %
CO	9.50	9.41	10.74	7.41	5.98	6.64	
CO_2	20.43	19.93	18.52	20.41	21.12	19.92	
CH_4	0.87	1.13	1.53	0.50	0.01	0.15	
N_2	48.56	52.75	56.78	56.29	56.73	61.75	
H_2/CO ratio	2.17	1.78	1.16	2.09	2.70	1.74	mol/mol
Lower heating value of dry obtained gas	3.74	3.41	3.25	2.78	2.50	2.14	MJ/Nm^3
Thermal capacity at yield of dry obtained gas	1.43	1.72	1.48	1.64	1.83	1.43	kW
Cold gas efficiency (without taking the capacity of the plasma torch into account)	67.14	61.43	58.04	59.35	50.97	44.00	%
Cold gas efficiency (taking the capacity of the plasma torch into account)	43.47	43.43	42.41	41.00	37.97	32.28	%
Hydrogen yield in terms of consumption of dry ash free (DAF) sewage sludge	0.76	0.53	0.46	0.61	0.67	0.49	Nm^3/kg

Table 7.3 The results of plasma steam-air gasification of mixture of sewage sludge pellets and rubber crumbs of worn tires (50/50 wt./wt.).

Parameter		Value		Unit
Consumption of dry mixture (moisture content is 3.9 wt. %)		0.29	0.31	kg/h
Thermal capacity by mixture consumption		1.41	1.51	kW
Electric capacity of steam plasma torch		1.28	1.36	kW
Water consumption in plasma torch		2.69	2.82	g/min
Air consumption		0.77	0.77	Nm^3/h
Oxygen consumption		0.16	0.16	Nm^3/h
Steam consumption		0.16	0.17	kg/h
Steam/oxygen ratio		1.24	1.30	mol/mol
The temperature of the outer walls of the reactor	Pyrolysis section	259	274	°C
	Oxidation section	416	432	
	Gasification section	443	413	
Yield of dry obtained gas		0.95	0.98	Nm^3/h
Composition of dry obtained gas				
H_2		10.98	12.70	vol. %
CO		8.05	8.78	
CO_2		16.10	15.55	
CH_4		0.73	0.56	
N_2		64.14	62.41	
H_2/CO ratio		1.36	1.45	mol/mol
Lower heating value of dry obtained gas		2.46	2.68	MJ/Nm^3
Thermal capacity at the yield of dry obtained gas		0.65	0.73	kW
Hydrogen yield in terms of consumption of dry ash free (DAF) sewage sludge		0.653	0.648	Nm^3/kg

Carbon conversion degree more than 90% and production of dry gas with hydrogen content of 45.3–64.9 vol. % were obtained in cases of using dry solid household waste and charcoal from corncob waste and the plasma torch electric capacity of 440–650 W up to 30 min of plasma-steam conversion.

Fast-growing algae carbon conversion degree of 97.6% and production of dry gas with hydrogen content of 65.4 vol. % were obtained at the plasma torch electric capacity of 990–1160 W in 19 min of plasma-steam conversion. The time of conversion and power of the plasma-steam torch affected the degree of carbon conversion of solid organic waste.

Sewage sludge carbon conversion degree of 39.8% and production of dry gas with hydrogen content of 71.8 vol. % were obtained at the plasma torch electric capacity of 500–520 W in 9 min of plasma-steam conversion.

In case of the use of charcoal from corncob waste, the degree of carbon conversion of 99.8% was achieved at the plasma torch capacity of 453–564 W in 30 min.

The equipment for plasma-steam gasification of solid organic material can be used for production not only for synthesis of gas, but also other useful products (charcoal, hydrogen).

The experimental plant for plasma steam-air gasification of solid organic waste with downdraft gasifier was created to increase the productivity of the plasma-steam installation with increasing rate of fixed carbon conversion using atmospheric oxygen and steam plasma torch Multiplaz 3500 tested under plasma-steam gasification.

The gas with H_2/CO molar ratio of 1.2–2.7 at a molar ratio of steam/oxygen in the blast of 0.5–1.1 was obtained during plasma steam-air gasification of wet sewage sludge in downdraft gasifier due to the use of a steam plasma torch. The yield of hydrogen was 0.5–0.8 N m^3/kg of dry ash free sewage sludge.

The gas with H_2/CO molar ratio of 1.4–1.5 at a molar ratio of steam/oxygen in the blast of 1.2–1.3 was obtained during plasma steam-air gasification of mixture of sewage sludge pellets and rubber crumbs of worn tires (50/50 wt./wt.) in downdraft gasifier. The yield of hydrogen was 0.65 Nm^3/kg of dry ash free waste mixture.

Obtained hydrogen-enriched gas can be used for the synthesis of valuable organic compounds.

It is planned to start research using oxygen-enriched air and a steam plasma torch for the operation of the downdraft gasifier.

ACKNOWLEDGMENTS

Scientific work in the framework of Target complex Programme for Research of NAS of Ukraine "Development of scientific bases for hydrogen production, storage and use in autonomous energy supply systems" was carried out. This work was supported by the Ministry of Education and Science of Ukraine under the program of the joint Ukrainian-Lithuanian R&D projects for the period of 2022–2023, Contract Number M/46–2022. The authors are grateful to the Ministry of Education and Science of Ukraine National Academy of Sciences of Ukraine for support of this research.

REFERENCES

Arsentiev I. V., Starik A. M., Zhovtyansky V. A. and Honcharuk Y. A. (2014). Nonequilibrium processes of nitric oxides formation in plasma-assisted waste gasification: modeling study. In: Advances in Nonequilibrium Processes: Plasma, Combustion, and Atmosphere, A. M. Starik and S. M. Frolov (eds), Torus Press, Moscow, pp. 272–277. ISBN: 978-5-94588-155-6.

Cai X. and Du C. (2021). Thermal plasma treatment of medical waste. *Plasma Chemistry and Plasma Processing*, **41**, 1–46, https://doi.org/10.1007/s11090-020-10119-6

Chernets O. V., Korzhyk V. M., Marynsky G. S., Petrov S. V. and Zhovtyansky V. A. (2008). Electric arc steam plasma conversion of medicine waste and carbon containing materials. Proc. of the XVII Int. Conf. on Gas Discharges and their Applications, September 7–12, Cardiff, Wales, pp. 465–468.

Diaz G., Leal-Quiros E., Smith R. A., Elliott J. and Unruh D. (2014). Syngas generation from organic waste with plasma steam reforming. *Journal of Physics: Conference Series*, **511**(012081), 1–6.

Dudnyk O. M. and Sokolovska I. S. (2010). Conversion of Ukrainian low grade solid fuels with CO_2 capture. Proc. of 27th Ann. Int. Pittsburgh Coal Conference 2010 (PCC 2010), October 11–14, Istanbul, Turkey, **2**, pp. 1012–1033.

Dudnyk O. M., Ostapchuk M. V., Sokolovska I. S. and Zhovtyansky V. A. (2025). Plasma steam-air conversion of solid biomass and problem of equilibrium assumption in the gasification process. *Energotehnologii I Resursosberezenie*, **84**(3), 3–20, https://doi.org/10.33070/etars.3.2025.01

Fabry F., Rehmet C., Rohani V.-J. and Fulcheri L. (2013). Waste gasification by thermal plasma: a review. *Waste and Biomass Valorization*, **4**(3), 421–439, https://doi.org/10.1007/s12649-013-9201-7

Favas J., Monteiro E. and Rouboa A. (2017). Hydrogen production using plasma gasification with steam injection. *International Journal of Hydrogen Energy*, **42**(16), 10997–11005, https://doi.org/10.1016/j.ijhydene.2017.03.109

Fernandes A., Brabandt J., Posdziech O., Saadabadi A., Recalde M., Fan I. L., Promes E. O., Liu M., Woudstra T. and Aravind P. V. (2018). Design, construction, and testing of a gasifier-specific solid oxide fuel cell system. *Energies*, **11**(8), 1985, https://doi.org/10.3390/en11081985

Lee H. G., Park H.-W., Choi S., Park H.-S. and Park D.-W. (2014). Production of synthesis gas from coal by DC non-transferred steam plasma gasification system. *Journal of Chemical Engineering of Japan*, **47**(4), 334–339, https://doi.org/10.1252/jcej.13we157

Materazzi M. and Taylor R. (2019). Plasma-assisted gasification for waste-to-fuels applications. *Industrial & Engineering Chemistry Research*, **58**(35), 15902–15913, https://doi.org/10.1021/acs.iecr.9b01239

Oliveira M., Ramos A., Ismail T. M., Monteiro E. and Rouboa A. (2022). A review on plasma gasification of solid residues: recent advances and developments. *Energies*, **15**(4), 1–21, https://doi.org/10.3390/en15041475

Pan P., Peng W., Li J., Chen H., Xu G. and Liu T. (2022). Design and evaluation of a conceptual waste-to-energy approach integrating plasma waste gasification with coal-fired power generation. *Energy*, **238**(C), 121947, https://doi.org/10.1016/j.energy.2021.121947

Petrov S. V. and Zhovtyansky V. A. (2019). Energy-efficient Steam-Plasma Technologies for Waste Processing. Naukova Dumka, Kyiv, Ukraine. ISBN 978-966-00-1683-5 (in Russian).

Serov A. A., Hrabovsky M., Kopecky V., Maslani A., Hlina M. and Hurba O. (2019). Lignite gasification in thermal steam plasma. *Plasma Chemistry and Plasma Processing*, **39**, 395–406, https://doi.org/10.1007/s11090-019-09957-w

Sikarwar V. S., Nageswara R. P., Vuppaladadiyam R. P., Krishna A., Ferreira N. L., Mašláni A., Tomar R., Pohořelý M., Meers E. and Jeremiáš M. (2022). Thermal plasma gasification of organic waste stream coupled with CO_2-sorption enhanced reforming employing different sorbents for enhanced hydrogen production. *RSC Advances*, **12**(10), 6122–6132, https://doi.org/10.1039/D1RA07719H

Singh H., Yadav R., Farooqui S. A., Dudnyk O. and Sinha A. K. (2019). Nanoporous nickel oxide catalyst with uniform Ni dispersion for enhanced hydrogen production from organic waste. *International Journal of Hydrogen Energy*, **44**(36), 19573–19584, https://doi.org/10.1016/j.ijhydene.2019.05.203

Tamošiūnas A., Valatkevičius P., Valinčius V. and Grigaitienė V. (2014). Production of synthesis gas from propane using thermal water vapor plasma. *International Journal of Hydrogen Energy*, **39**(5), 2078–2086, https://doi.org/10.1016/j.ijhydene.2013.11.134

Uhm H. S., Na Y. H., Hong Y. C., Shin D. H. and Cho C. H. (2014). Production of hydrogen-rich synthetic gas from low-grade coals by microwave steam-plasmas.

International Journal of Hydrogen Energy, **39**(9), 4351–4355, https://doi.org/10.1016/j.ijhydene.2014.01.020

Zhang Q., Dor L., Fenigshtein D., Yang W. and Blasiak W. (2012). Gasification of municipal solid waste in the plasma gasification melting process. *Applied Energy,* **90**(1), 106–112, https://doi.org/10.1016/j.apenergy.2011.01.041

Zhovtyansky V. A. and Anisimova O. V. (2014). Kinetics of plasma chemical reactions of producing nitrogen atoms in the glow discharge in a nitrogen-argon gas mixture. *Ukrainian Journal of Physics,* **59**(6), 1155–1163, https://doi.org/10.15407/ujpe59.12.1155

Zhovtyansky V. and Valinčius V. (2018). Efficiency of plasma gasification technologies for hazardous waste treatment. In: Gasification for Low-Grade Feedstock, Y. Yun (ed.), IntechOpen, London, United Kingdom of Great Britain, pp. 65–189.

Zhovtyansky V. and Ostapchuk M. (2022). Plasma technologies in the problem of obtaining "more than green hydrogen". *Combustion and Plasma Chemistry,* **20**(1), 11–32 (in Russian).

Zhovtianskyi V., Dudnyk O., Nevzgliad I. and Sokolovska I. (2011). Hydrogen rich gas generation using plasma steam gasification of Ukrainian anthracite and brown coal. Proc. of Int. Conf. on Hydrogen Production ICH_2P-11, June 19–22, Thessaloniki, 246, pp. 1–9.

Zhovtyansky V., Dudnyk O., Petrov S., Verbovsky V., Rubets D. and Yakymovych M. (2013a). Plasma-steam intensification of the hydrogen yield in the process of wood air gasification. *High Temperature Material Processes–An International Quarterly of High-Technology Plasma Processes,* **17**(1), 25–35, https://doi.org/10.1615/HighTempMatProc.2014012582

Zhovtyansky V. A., Petrov S. V., Lelyukh Y. I., Nevzglyad I. O. and Goncharuk Y. A. (2013b). Efficiency of renewable organic raw materials conversion using plasma technology. *IEEE Transactions on Plasma Science,* **41**(12), 3233–3239, https://doi.org/10.1109/TPS.2013.2275936

Zhovtyansky V., Kolesnikova E., Yakymovych M. and Seredenko P. (2018). The general principles of waste processing with recovery of their energy potential on the basis of plasma technologies. Part III. Comparative analysis of the oxygen and air blowing influence and the role of calorific content of sewage sludge. *Energy Technologies & Resource Saving,* **2**, 16–30 (in Ukrainian), https://doi.org/10.33070/etars.2.2018.03

Zhovtyansky V. A., Dudnyk O. M., Ostapchuk M. V. and Sokolovska I. S. (2023). Conversion of carbon-containing raw material using plasma technologies. *Energotehnologii I Resursosberezenie,* **77**(4), 90–112, https://doi.org/10.33070/etars.4.2023.08

doi: 10.2166/9781789065138_0097

Chapter 8

Utilizing waste fine granite to produce sustainable dry pressed ceramic tiles

*Sh. K. Amin[1], N. Y. S. Selem[2] and N. F. Abdel Salam[1]**

[1]Chemical Engineering and Pilot Plant Department, Engineering and Renewable Energy Research Institute, National Research Centre (NRC), Giza, Egypt
[2]Higher Technological Institute, 10th of Ramadan City, Egypt
[3]Chemical Engineering Department, Faculty of Engineering, Cairo University, Giza, Egypt
**Corresponding author: nourafathy@eng.cu.edu.eg*

ABSTRACT

The need for building materials has increased as a result of population growth and fast urbanization, which require the construction of more residential units. In this study, clay has been substituted with waste materials, which has helped to reduce natural clay consumption. Investigating the characteristics of sustainable dry-pressed ceramic tiles made by adding fine granite waste (FGW) such as the modulus of rapture (MOR), bending strength , and water absorption percentage (WA%) were the main areas of study. Ceramic floor and wall tiles were created by substituting fine granite waste (FGW) in four different mixtures (0% and 5%, 10% and 15% FGW) for the commercial primary mix of tiles. Regarding ceramic wall tiles, WA met the standards for all waste percentages, MOR did not meet the standards for all waste percentages, and BS met the required standards at 5% replacement only. Therefore, 5% additive was the best outcome in that scenario because it reduced water absorption and met the breaking strength (BS) requirement. Whereas WA met the requirements for floor ceramic tiles at waste percentages of 10% and 15%, MOR and BS met the requirements for all waste percentages. Therefore, 15% FGW was the best outcome in that situation since it would enable clay-based materials to be sustainable by using inexpensive resources (waste materials), which will lessen the environmental problems brought on by waste disposal.

Key words: Ceramic tiles, clay, granite, sustainability, wastes

8.1 INTRODUCTION

The demand for construction materials has expanded as a result of population growth and rapid urbanization (Churkina *et al.*, 2020; Nayak *et al.*, 2022). The most common building material that is collected from top soil for making clay bricks and clay tiles is clay, which has been the case for decades. However, clay extraction depletes natural clay resources, which are virgin soils that can be utilized for farming and are naturally fertilized (Kazmi *et al.*, 2016b).

There is an urgent need to find natural or artificial clay substitutes due to rising use and restrictions on the extraction of natural clay from the earth. Utilizing waste materials instead of clay reduces the need for natural clay and addresses environmental problems brought on by trash disposal, especially that Egypt produces about 90.76 million tons solid waste annually (Daoud *et al.*, 2020; Subashi De Silva *et al.*, 2022).

It has been investigated to substitute clay with waste materials by many researchers. For example, industrial ceramic sludge (Coletti *et al.*, 2016), paper sludge (Cusidó *et al.*, 2015; Vieira *et al.*, 2016), waste rice husk (Kazmi *et al.*, 2016a), waste sugarcane bagasse (Kazmi *et al.*, 2016b), marble and granite waste (Hamza *et al.*, 2011) were used in brick production with different percentage to replace clay.

In the manufacturing of floor tiles, clay also was substituted by sewage sludge by 7% (Amin *et al.*, 2018).

While, rice husk ash by 10% (De Silva & Surangi, 2017), ceramic sludge by 20% (Subashi De Silva & Mallwattha, 2018), glass waste by 10% (Costa *et al.*, 2009), granite waste and coffee husk ash (Acchar *et al.*, 2016) in the production of roof tiles.

Ceramic tiles are widely used because of their great strength, wear, and stain resistance, as well as their low water absorption (Wen *et al.*, 2023).

For walls, roofs, and floors that are frequently used in construction, ceramic tiles have emerged as the most popular decorative materials (Elakhame *et al.*, 2016). Ceramic tiles have replaced painting as the most prevalent building materials as a result of substantial technological advancement in recent decades, which has significantly raised the standard of living and infrastructure in developing nations (Ngayakamo *et al.*, 2022).

As a result, there is currently a significant growth in the demand for ceramic tiles for construction projects, and this demand is expected to continue to grow in the future (Hossain & Roy, 2020). On the other hand, the significant expansion of the ceramics industry has led to a huge intake of clay raw materials, their overuse, and ultimately negative environmental effects (Hossain & Roy, 2020; Muthukannan & Ganesh, 2019). For instance, the manufacturing of ceramic tiles necessitates a high flux consumption of about 50–60% by weight (Kara *et al.*, 2009). As a result, it is crucial to recycle and repurpose industrial waste into useful resources in accordance with the circular economy.

At the same time, one of the most prevalent industrial by-products discharged from the construction sectors is granite waste. Despite having special chemical qualities, granite garbage is nevertheless wasted in large quantities without being recycled, which could have a negative influence on the environment (Ngayakamo *et al.*, 2022).

Due to the presence of alkaline feldspar (30–60%) and quartz (10–40%) in granite waste, it can be used as feldspar (Shamsabadi *et al.*, 2018). Due to the presence of K_2O and Na_2O, two necessary fluxing agents, granite powder may be considered as a potential alternative flux material (Hojamberdiev *et al.*, 2011).

The processing of granite rock results in the production of granite waste, which is then released into the environment as micronized stones, sludge, or powder (Amin *et al.*, 2020; Ostrowski *et al.*, 2020). Granite debris has subsequently been labelled as an airborne and toxic material that contaminates underground water, decreases soil porosity, causes lung and nasal ailments, and halts the natural flow of aquifers (Ghannam *et al.*, 2016; Shamsabadi *et al.*, 2018).

Due to the damaging environmental effects of granite waste, it is necessary to recycle it and turn it into a value-added product in accordance with the circular economy to lessen its influence on the environment (Kothari *et al.*, 2021). So, this study offers a workable strategy for recovering and using granite waste as substitution of clay in the manufacture of ceramic tiles.

8.2 METHODS

8.2.1 Raw materials

The materials used in this study were obtained from the local Egyptian market.

Ceramic wall and floor mixtures were obtained directly as powder from 'Ceramica Venus' factory, 10th of Ramadan city–Egypt. These conventional mixtures were prepared from Egyptian raw materials such as ball and Aswan clays from Aswan city, potash feldspar from Eastern desert, bentonite from Borg El–Arab, Alexandria, Glass sand (Quartz) from Zaafarana, Suez Gulf and limestone from El-Menia.

The fine granite waste (FGW) powder was obtained from Shaqu-Elteban area, Egypt. This fine waste was used to substitute conventional mixture.

Average bulk density of floor and wall mixes powder, and granite waste was 1.79, 1.73, and 2.31 gm/cm^3, respectively.

8.2.2 Materials characterization

8.2.2.1 Chemical analysis (XRF)

X-ray fluorescence spectrometry (XRFS) is a method of elemental analysis that assesses the presence and concentration of various elements by measurement of secondary X-radiation from the sample that has been excited by an X-ray source. The instrument XRF analysis is run on an AXIOS, PANalytical 2005, wavelength dispersive sequential spectrometer (WD–XRF), which is installed at the National Research Centre (NRC) (Sikalidis 2011).

8.2.2.2 Mineralogical analysis (XRD)

XRD analyzed by a BRUKUR D8 ADVANCEDCOMPUTERIZED X–Ray Diffractometer apparatus (At the Center Metallurgical Research and Development Institute) with mono–chromatized Cu Kα radiation, operated at 40 kV and 40 mA (Sikalidis 2011).

8.2.2.3 Thermal analysis (DTA and TGA)

DTA and TGA were recorded on equipment "NETZSCH STA 409C/CD" (Installed at Center Metallurgical Research and Development Institute).

8.2.2.4 Screen analysis

For the determination of the grain-size distribution, the standard sieving procedure described by ASTM D 422/2007 (ASTM International 2022b) has been used. Also, the sieves used are in compliance with ASTM E 11/2020 (ASTM International 2022c).

8.2.3 Sample preparation

Floor and wall ceramic tiles were formed by replacing the commercial powder of tiles by fine granite waste (FGW) in four mixtures: 0% and 5%, 10% and 15% FGW.. The prepared samples were molded by dry pressed in a uniaxial hydraulic press under a pressure of 10 MPa in $100 \times 25 \times 15$ mm^3 molds. Finally, the tile specimens were fired using the laboratory furnace, Protherm–electrical furnace model PLF 14015, for 15 min soaking time. The maximum firing temperature is (1050°C) for wall samples, and (1150°), for floor samples. Heating rates were chosen to be as close as possible to industrial conditions. The firing technique used in this work is the single fast firing using the laboratory box furnace.

8.2.4 Water absorption percentage (WA%) test

Water absorption is a critical physical performance indicator for determining the quality of ceramics. Water absorption is commonly used to reflect the degree of sintering of ceramic products and to indirectly indicate the size of apparent porosity. It is also regarded as a significant indicator of the level of vitrification/densification. According to ASTM C373 (ASTM International 2022a), for the determination of vitrification parameters, the fired specimens were individually tested as follows in the chemical laboratory, Chemical Engineering Department, Faculty of Engineering, Cairo University

8.2.5 Modulus of rapture test

The modulus of rapture (MOR) test is the standard for determining the mechanical properties of ceramics. MOR testing machines are generally used for testing the strength of ceramic tiles and other materials.

MOR of fired specimens at different percent waste addition, were determined according to international standard ISO 10545-4 (ISO the International Organization for Standardization, 2019).

8.2.6 Breaking strength (BS) test

BS is the ability of a material to withstand a pulling or tensile force. It is customarily measured in units of force per cross-sectional area. The purpose of this test is to determine at what point a tile will break when a load is applied to a specific point on the tile. BS of fired specimens at different percent waste addition, were determined according to international standard ISO 10545-4 (ISO the International Organization for Standardization, 2019).

8.3 RESULTS AND DISCUSSION

8.3.1 Chemical analysis (XRF)

The XRF analysis of floor and wall mixes powder, and granite waste is shown in Table 8.1. Granite waste reveals the predominance of silica in a proportion exceeding 60%.

Table 8.1 Chemical (XRF) analysis of raw materials.

Main Constituents	Granite Waste	Wall Mixture	Floor Mixture
SiO_2	67.85	55.51	61.21
TiO_2	0.44	0.91	0.83
AlO_3	15.68	19.73	20.19
Fe_2O_3	3.13	5.10	4.78
MgO	0.63	0.40	0.99
CaO	1.43	5.15	1.21
Na_2O	4.51	1.43	2.72
K_2O	4.82	1.17	1.21
P_2O_5	0.11	0.23	0.21
SO_3	0.09	0.31	0.28
Cl	0.05	0.09	0.10
Minor oxides	0.43	0.28	0.28
LoI	0.77	9.68	5.98
Total	99.94	99.99	99.99

8.3.2 Mineralogical analysis of raw materials (XRD)

Mineralogical analysis of raw materials shown in Figure 8.1 display that both wall and floor tiles mixes are composed of quartz, kaolinite, albite. Wall tiles XRD further exhibits lines of calcite ($CaCO_3$) while these do not appear on the XRD pattern of floor tiles due to their limited percentage.

On the other hand, the granite waste is composed of: quartz (SiO_2), microcline ($K_2O.Al_2O_3.6SiO_2$), and albite ($Na_2O.Al_2O_3.6SiO_2$).

8.3.3 Combined TGA–DTA charts

Combined TGA–DTA charts for tile mixes and FGW are shown in Figure 8.2. It appears from both charts of tile mixes that there is a slight early decrease in weight due to elimination of physical water followed by a diffuse exothermic peak ending at about 400°C due to oxidation of organic impurities. An endothermic peak follows ending at about 525° presumably due to loss of lattice water of clays present in the mixes that is practically completed at about 650°C according to the TGA trace. In case of wall tiles and owing to the presence of limestone a supplementary peak can be observed at about 720°C.

The thermal decomposition of granite shows practically no weight loss (less than 1% at 1000°C) reflecting the absence of any decomposable salts in its composition.

8.3.4 Particle-size analysis

The particle-size distribution of the three raw materials was determined by the standard sieves and is shown in Figure 8.3. The cumulative analysis is presented in Figure 8.3 and reveals that the particle-size distributes between 0.1 and 1 mm.

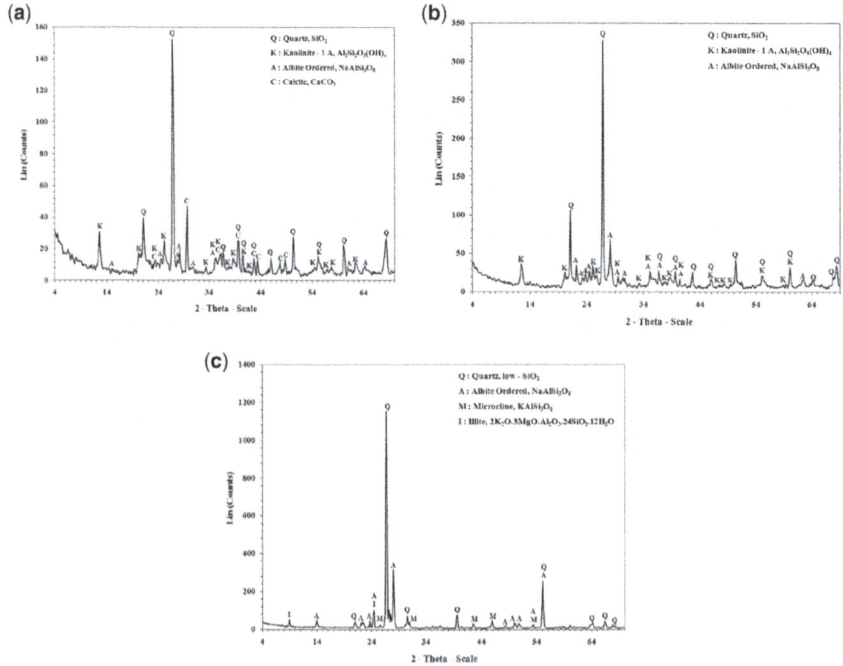

Figure 8.1 (a) XRD pattern of wall Mix, (b) XRD pattern of floor Mix, (c) XRD pattern of granite waste.

8.3.5 Water absorption test results

The percent water absorption results are shown in Figure 8.4 for wall and floor tiles with different percentage of FGW. The wall tiles standard water absorption value is 10–20% while that of floor tiles is 6–10% (ISO the International Organization for Standardization 2018). Here it is found that the water absorption percentage is less than 20% but it is still high percentage compared with the standard limits for wall tiles.

For floor tiles, it is noticed that all percentages of FGW met the required standard and even below the minimum value which show promising results.

8.3.6 Modulus of rapture test results

The MOR results are shown in Figure 8.5 for wall and floor tiles with different percentage of FGW. The wall tiles standard MOR has to be more than 12 N/mm^2 while the minimum value of MOR for floor tiles is 18 N/mm^2 (ISO the International Organization for Standardization, 2018). Here it was found that wall tiles MOR results did not meet the required specifications while that of floor tiles were more than the minimum required value for all FGW percentages.

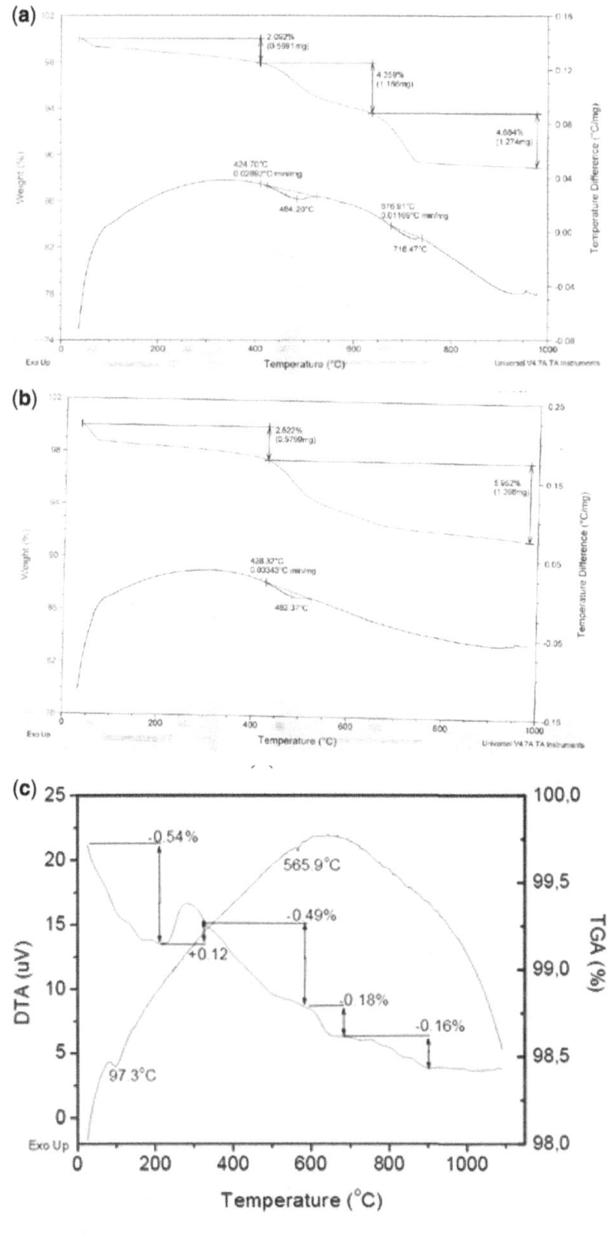

Figure 8.2 (a) TGA and DTA pattern of wall Mix, (b) TGA and DTA pattern of floor Mix, (c) TGA and DTA pattern of granite waste.

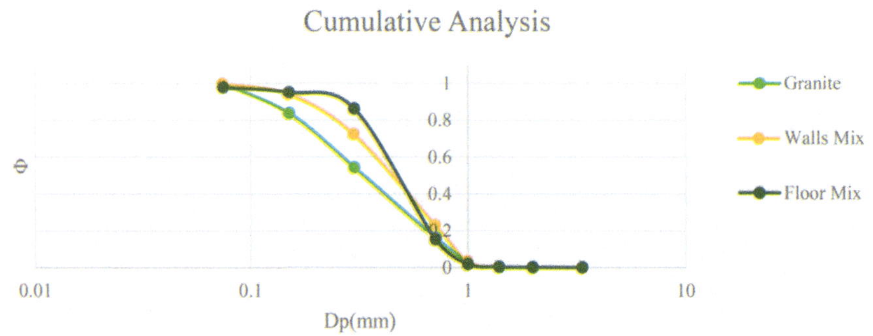

Figure 8.3 Screen analysis.

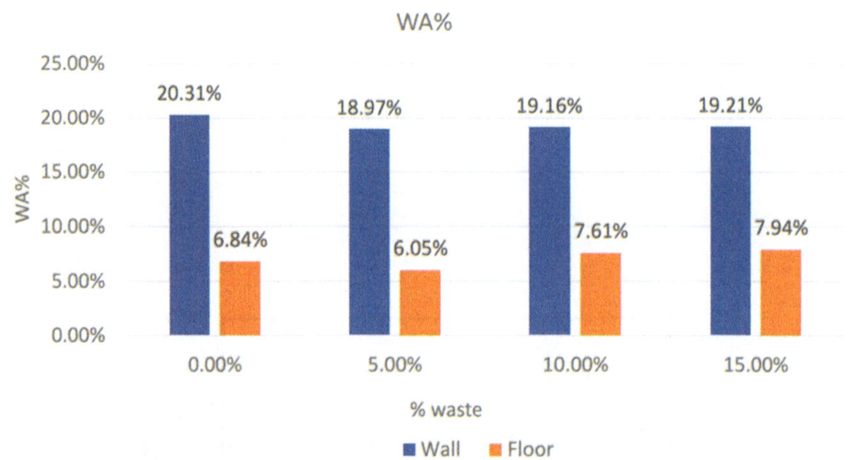

Figure 8.4 Water absorption test results.

8.3.7 Breaking strength test results

The BS results are shown in Figure 8.6 for wall and floor tiles with different percentage of FGW. The wall tiles standard BS has to be more than 600 N while that of floor tiles has to be more than 800 N (ISO the International Organization for Standardization, 2018). Here it was found that wall tiles BS results did not meet the required specifications except for 5% FGW as additive while that of floor tiles were more than the minimum required value for all FGW percentages.

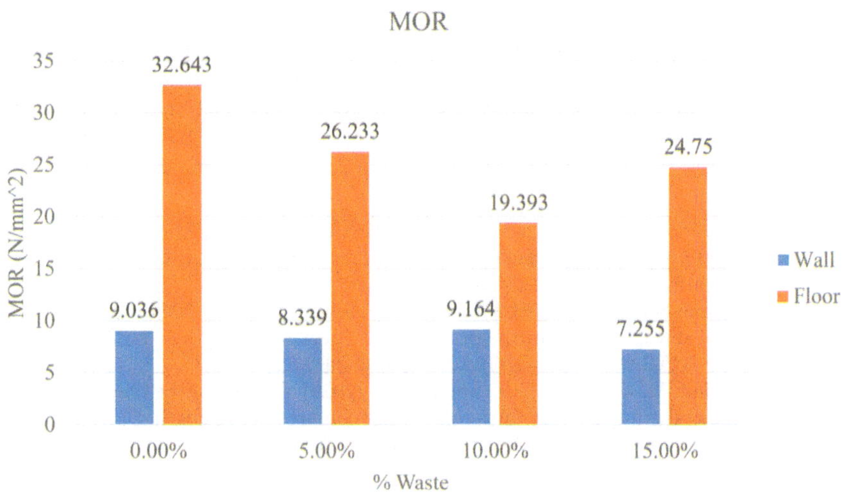

Figure 8.5 MOR test results.

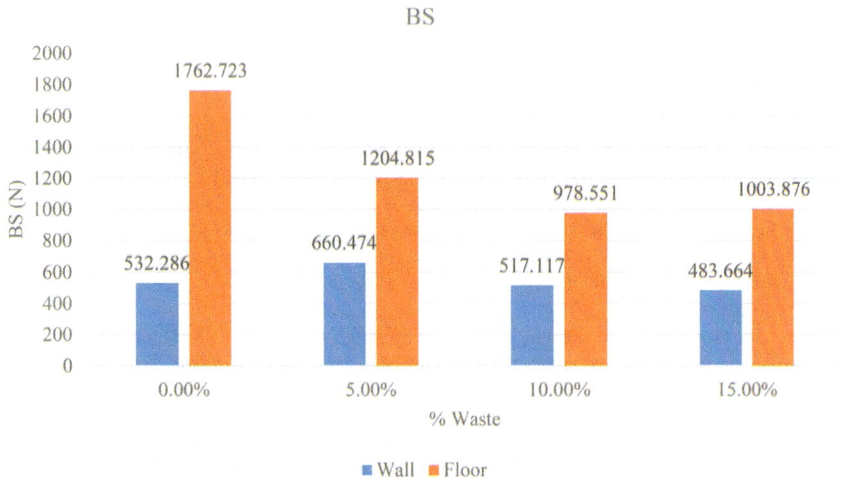

Figure 8.6 BS test results.

8.4 CONCLUSIONS

In case of wall ceramic tiles, WA achieved the standards for all percentages of waste, MOR did not achieve the standards for all percentages of waste, and BS achieved the standards at 5% waste only. So, the optimum result in that case was 5% FGW due to lower water absorption and achieving the standard of BS.

While in case of floor ceramic tiles, WA achieved the standards at percentages of 10% and 15% waste, MOR and BS achieved the standards for all percentages of waste. So, the optimum result in that case was 15% FGW which will achieve sustainability of clay-based materials by utilizing low-cost materials (waste materials), which will reduce the environmental issues caused by waste disposal.

REFERENCES

Acchar W., Avelino K. A. and Segadães A. M. (2016). Granite waste and coffee husk ash synergistic effect on clay-based ceramics. *Advances in Applied Ceramics*, **115**(4), 236–242, https://doi.org/10.1080/17436753.2015.1126989

Amin S. K., Abdel Hamid E. M., El-Sherbiny S. A., Sibak H. A. and Abadir M. F. (2018). The use of sewage sludge in the production of ceramic floor tiles. *HBRC Journal*, **14**(3), 309–315, https://doi.org/10.1016/j.hbrcj.2017.02.002

Amin S. K., Allam M. E., Garas G. L. and Ezz H. (2020). A study of the chemical effect of marble and granite slurry on green mortar compressive strength. *Bulletin of the National Research Centre*, **44**(1), https://doi.org/10.1186/s42269-020-0274-8

ASTM International (2022a). 'ASTM C373-17: Standard Test Methods for Determination of Water Absorption and Associated Properties by Vacuum Method for Pressed Ceramic Tiles and Glass Tiles and Boil Method for Extruded Ceramic Tiles and Non-tile Fired Ceramic Whiteware Products,' in ASTM International.

ASTM International (2022b). 'ASTM D 422/1963 (Reapproved 2007), 'Method for particle–size analysis of soils', Annual book of American Society for Testing of Material (ASTM), U.S.A., 4 (8)'.

ASTM International (2022c). 'ASTM E 11/2020, 'Specifications for wire–cloth sieves for testing purposes', Annual book of American Society for Testing of Material (ASTM), U.S.A., 14 (2)'.

Churkina G., Organschi A., Reyer C. P. O., Ruff A., Vinke K., Liu Z., Reck B. K., Graedel T. E. and Schellnhuber H. J. (2020). Buildings as a global carbon sink. *Nature Sustainability*, **3**(4), 269–276, https://doi.org/10.1038/s41893-019-0462-4

Coletti C., Maritan L., Cultrone G. and Mazzoli C. (2016). Use of industrial ceramic sludge in brick production: effect on aesthetic quality and physical properties. *Construction and Building Materials*, **124**, 219–227, https://doi.org/10.1016/j.conbuildmat.2016.07.096

Costa F. B., Teixeira S. R., Souza A. E. and Santos G. T. A. (2009). Recycling of glass cullet as aggregate for clays used to produce roof tiles. *Matéria (Rio de Janeiro)*, **14**(4), 1146–1153, https://doi.org/10.1590/s1517-70762009000400007

Cusidó J. A., Cremades L. V., Soriano C. and Devant M. (2015). Incorporation of paper sludge in clay brick formulation: ten years of industrial experience. *Applied Clay Science*, **108**, 191–198, https://doi.org/10.1016/j.clay.2015.02.027

Daoud A. O., Othman A. A. E., Robinson H. and Bayyati A. (2020). An investigation into solid waste problem in the Egyptian construction industry: a mini-review. *Waste Management & Research: The Journal for A Sustainable Circular Economy*, **38**(4), 371–382, https://doi.org/10.1177/0734242X20901568

De Silva G. H. M. J. S. and Surangi M. L. C. (2017). Effect of waste rice husk ash on structural, thermal and run-off properties of clay roof tiles. *Construction and Building Materials*, **154**, 251–257, https://doi.org/10.1016/j.conbuildmat.2017.07.169

Elakhame Z. U., Ifebhor F. and Asotah W. A. (2016). Development and production of ceramic tiles from waste bottle powder (milled glass). *Kathmandu University Journal of Science, Engineering and Technology*, **12**(2), 50–59, https://doi.org/10.3126/kuset.v12i2.21521

Ghannam S., Najm H. and Vasconez R. (2016). Experimental study of concrete made with granite and iron powders as partial replacement of sand. *Sustainable Materials and Technologies*, **9**, 1–9, https://doi.org/10.1016/j.susmat.2016.06.001

Hamza R. A., El-Haggar S. and Khedr S. (2011). Utilization of marble and granite waste in concrete bricks. *International Journal of Bioscience and Biochemistry*, **21**.

Hojamberdiev M., Eminov A. and Xu Y. (2011). Utilization of muscovite granite waste in the manufacture of ceramic tiles. *Ceramics International*, **37**(3), 871–876, https://doi.org/10.1016/j.ceramint.2010.10.032

Hossain S. S. and Roy P. K. (2020). Sustainable ceramics derived from solid wastes: a review. *Journal of Asian Ceramic Societies*, **8**(4), 984–1009, https://doi.org/10.1080/21870764.2020.1815348

ISO (the International Organization for Standardization) (2018). '13006:2018(en), ISO, Ceramic tiles–Definitions, classification, characteristics and marking,' [Online]. Available: https://www.iso.org/standard/63406.html

ISO (the International Organization for Standardization) (2019). '10545-4:2019(en), ISO, Ceramic tiles – Part 4: Determination of modulus of rupture and breaking strength', [Online]. Available: https://www.iso.org/obp/ui/#iso:std:iso:10545:-4:ed-4:v1:en

Kara A., Kayaci K., Küçüker A. S., Bozkurt V., Üçbas Y. and Özdamar S. (2009). Use of rhyolite as flux in porcelain tile production. *Indian Ceramics*, **29**(2).

Kazmi S. M. S., Abbas S., Munir M. J. and Khitab A. (2016a). Exploratory study on the effect of waste rice husk and sugarcane bagasse ashes in burnt clay bricks. *Journal of Building Engineering*, **7**, 372–378, https://doi.org/10.1016/j.jobe.2016.08.001

Kazmi S. M. S., Abbas S., Saleem M. A., Munir M. J. and Khitab A. (2016b). Manufacturing of sustainable clay bricks: utilization of waste sugarcane bagasse and rice husk ashes. *Construction and Building Materials*, **120**, 29–41, https://doi.org/10.1016/j.conbuildmat.2016.05.084

Kothari R., Sahab S., Singh H. M., Singh R. P., Singh B., Pathania D., Singh A., Yadav S., Allen T., Singh S. and Tyagi V. V. (2021). COVID-19 and waste management in Indian scenario: challenges and possible solutions. *Environmental Science and Pollution Research*, **28**(38), 52702–52723, https://doi.org/10.1007/s11356-021-15028-5.

Muthukannan M. and Ganesh A. S. C. (2019). The environmental impact caused by the ceramic industries and assessment methodologies. *International Journal for Quality Research*, **13**(2), 315–334, https://doi.org/10.24874/IJQR13.02-05

Nayak S. K., Satapathy A. and Mantry S. (2022). Use of waste marble and granite dust in structural applications: a review. *Journal of Building Engineering*, **46**, 103742, https://doi.org/10.1016/j.jobe.2021.103742

Ngayakamo B., Bello A. and Onwualu A. P. (2022). Valorization of granite waste powder as a secondary flux material for sustainable production of ceramic tiles. *Cleaner Materials*, **4**, 100055, https://doi.org/10.1016/j.clema.2022.100055

Ostrowski K., Stefaniuk D., Sadowski Ł, Krzywiński K., Gicala M. and Różańska M. (2020). Potential use of granite waste sourced from rock processing for the application as coarse aggregate in high-performance self-compacting concrete. *Construction and Building Materials*, **238**, 117794, https://doi.org/10.1016/j.conbuildmat.2019.117794

Shamsabadi E. A., Ghalehnovi M., de Brito J. and Khodabakhshian A. (2018). Performance of concrete with waste granite powder: the effect of superplasticizers. *Applied Sciences*, **8**(10), 1808, https://doi.org/10.3390/app8101808

Sikalidis C. (ed.) (2011). Advances in Ceramics–Characterization, Raw Materials, Processing, Properties, Degradation and Healing. InTech.

Subashi De Silva G. H. M. J. and Mallwattha M. P. D. P. (2018). Strength, durability, thermal and run-off properties of fired clay roof tiles incorporated with ceramic sludge. *Construction and Building Materials*, **179**, 390–399, https://doi.org/10.1016/j.conbuildmat.2018.05.187

Subashi De Silva G. H. M. J., Aagani T. H. F., Gebremariam K. F. and Samarakoon S. M. S. M. K. (2022). Engineering properties and microstructure of a sustainable roof tile manufactured with waste rice husk ash and ceramic sludge addition. *Case Studies in Construction Materials*, **17**, e01470, https://doi.org/10.1016/j.cscm.2022.e01470

Vieira C. M. F., Pinheiro R. M., Rodriguez R. J. S., Candido V. S. and Monteiro S. N. (2016). Clay bricks added with effluent sludge from paper industry: technical, economical and environmental benefits. *Applied Clay Science*, **132–133**, 753–759, https://doi.org/10.1016/j.clay.2016.07.001

Wen L., Lin L., Fan Y., Luo Y., Ma S., Zhou Y., Yang C., Shih K. and Li X. (2023). Valorization of thermally hydrolyzed sludge with clay for sintering of ceramic tiles. *Science of the Total Environment*, **877**, 162871, https://doi.org/10.1016/j.scitotenv.2023.162871

Index

H

I

L

M

N

O

P

Q

R

S

IWA Publishing's authorised EU representative for General Product Safety Regulations is Diane D'Arras, 15 rue Duret, 75116 Paris, France, e-mail: safety@iwap.co.uk.

Printed and bound by CPI Group (UK) Ltd, Croydon, CR0 4YY
27/03/2026
02079976-0003